Pipeline Diaries

Author: Ryan Wiersma

Editor: Dallas Garhardt

Blog: www.albertaforever.com

Credit to Charles Bukowski

Big Thanks

I want to send out a huge thanks to my friend and editor, Dallas Garhardt, who helped edit and guide me throughout this entire book. Much love sent your way!

A big thanks to Brayden Kagel, who was a great friend in giving me encouragement and always giving an honest opinion on how the book could be better. Thank you so much!

To all the lunatics who provided the inspiration
to make this book happen, thank you.
Life would be boring without you.
It will be a glorious day
when I see you bastards in hell!

Introduction

We are stone, pure madness. We are the people who belong in loony bins, jailhouses, back alleys—anywhere you can stop for a piss on the road straight to hell.

Have you ever felt a pull so strong it yanks you away from who you thought you were? It did to who I thought I was.

A job haggles away your soul: your essence. Gets sold for thirty-five-thousand dollars a year, plus dental, if you're lucky.

I never fit behind a desk. None of these boys ever felt comfortable inside walls.

Rig pigs. Dirty, filthy pipeliners. We descend from the patch like the harbingers of the apocalypse in jacked-up F350s, rolling coal, black smoke darker than crude oil itself. You hate us for that—bitching about us behind the wheel of your Toyota Camry, making jokes about the size of our unit. Is it because we can get what we want and do what we want?

Hide your children, hide your daughters, hide your wives.

Most of our crew are in broken homes, or from

busted up relationships, divorced by beautiful, beautiful gold diggers. Whore lovers and beer chugging madmen get together when excess loves company. We are the boys… Lavish, crazy, glorious…

Twelve hour days are nothing. I really mean that. Fuck, give us a few rips and we'll go for days. Some of the best guys I know are out here. Loyal. I mean, really loyal. Out here, that's all you got. When it's forty below and you're broken down, and out, and done, and you still have to get the work finished, you're thanking the heavens that you have your buds with you.

When I'm sure I'm going to die, when my fingers are blue, useless sticks, and I can't feel my feet through steel toes, because I'm selling my soul for someone else to sell a barrel of black gold, I still know I'll do it over and over again …

When I first came up here, before they beat the boy out of me, I never thought I'd say it, but it's a wonderful thing to see suffering. It's even better to be a part of it, tasting it while it slaps me around and calls me worthless.

It's okay to call us God. You can do that. Call us a dream. Or maybe call us truth, because you only find truth through pain, can only reveal a man's character through a fight. And if you cry, you might as well leave, because we don’t tolerate the weak. Junkyard dogs don’t play well with little bitches.

We have a saying, and it reads like gospel from the Bible itself:

You can’t be a pussy and a pipeliner.

The Dream

"The dreamers of the day are dangerous men, for they may act on their dreams with open eyes, to make them possible."
- T.E. Lawrence

I have a dream, and it starts in Ecuador.

I don't think I'm alone in believing that things don't happen *to* you; they happen *because* of you. You lost your job, because you didn't work hard enough. You feel lonely, because you're alone, because you have deep flaws. You didn't realize your dream, because you don't deserve a happy ending.

I've never been much of a sleeper. If you go to bed, you might as well go to work—it's a long blink away from needing to drag yourself out from under blankets to be tackled by another day. So, I stay awake to spite the night, dreaming in the reality of the day when the dream can be controlled, made my own through planning, sweat, and blood.

The dream, it's like Rome—so fragile, so

delicate. That's why I hold tight and stay vigilant, careful not to get knocked off balance by life's pushes and pulls. The slightest breeze, and it would be gone, pissed away to ash. Nothing would remain except a brief memory of a faint glimmer of the hope that once drove me.

When I walk the streets in a city—Edmonton, Hong Kong, Sau Paulo, Bangkok, it doesn't matter—I try to meet eyes with as many people as I can. I want to see if they still have that spark. You can tell when someone doesn't. You can just tell. The colour seems drained, gray and waiting, like the sky before a faraway storm.

Family and teachers are always about the safe way, the way that makes the cautious cash. Listen to your parents; they know best. Keep your head down. Go to school to be a teacher, or a doctor, or a mechanic. Take business. Get a job that they understand. That's the safe way.

Fuck your childhood dream; you're not a child anymore.

Would you want to be a child again? Roll the dice on another version of yourself? Would you break the same way or in some wonderful new way that made you more interesting at parties?

To tell you the truth, I'm not sure what I would even change. I had a good childhood. I lived in an average home with good parents; a loving mother and a good father, a true man who taught me to be respectful to people, say "please" and "thank you." I was taught not to fight and not to yell unless it was to shout *no* to drugs.

It's weird how normal it must feel to raise the perfect breed of sheep. An easy to control taxpayer who fears the law and apologizes when wronged to avoid anything resembling confrontation. Why is developing resentment to this not normal? Being against all that shit seems like the most normal thing I could imagine. I did what many do: I left to travel the world, search for truth in beaches, and find simplicity in crowded street markets.

I liked the heat. I would sweat and sweat and feel cleansed, baptized within myself. I liked the third world country way—laidback vibes and a chilled attitude towards life. A deep bond to their community, love in their hearts. It was refreshingly addictive. They were real. They knew pain. They knew suffering and hunger. And yet, people were always smiling with eyes still bright, alive with dreams.

I didn't know pain, not really. I didn't know

suffering or hunger. I didn't know shit, but I knew enough to realize that real wisdom can come from knowing you don't know jack shit about real life.

It was weird to think of back home and see how spoiled kids that were given everything—H2 Hummers bought by daddy to try to get them out of the basement where they dwelled, just finger-banging games on their phone—were all unhappy. A whole generation of soft, spoiled children knowing nothing of pain or hardship, death or hunger, but this consistent, humming sadness all the same. I could see there was a sickness in my country. There was an illness in Canada, in the whole western world. I decided I needed a new home.

I bounced around the world for many years, trying every country like it was a different flavour in a box of chocolates. I've tasted the Maori Haka Dance of New Zealand, delighted in the beaches of Australia, pedal biked across Southeast Asia into China, bathed in the righteous, exotic kindness of Brazil during the World Cup, and found myself in a little town in Ecuador called Montanita. There, I surfed epic waves before lying on the beach to drink cold beer and view all the talent in bikinis. Ecuador! This is the place that finally felt right. Chill during the day and party during the night! You can think bigger here, let your mind run on more important things, not like what we'll watch next on Netflix, or when we'll see how our friend rearranged her cat's toys.

I took the leap and bought a piece of land in the country of Ecuador. All the money I had at the time was now gone. Sixty-thousand American dollars, and I was the proud owner of half an acre of ocean view land in the little party town of Montanita. Just the land: no roads, no water, no electricity. I needed more money! And I needed it fast!

I'm from Alberta, and Alberta has oil, and oil spells money. Black gold, some say. Exactly what I needed. Like I said, I have dream. I will achieve it, or I will die trying.

My Heart Ripped Out

I worked my hands to the bone, laying pipe in the barren cold for three seasons, twenty-four months total, only to spend all the money I made all at once. But I was the proud, very proud, owner of half an acre. I knew my future was more Alberta, plundering oil from the earth so some other guy can get rich. But, for now, I was in Ecuador. I had made it. And I was to build a cathedral atop this humble foundation.

I was going to have a house in paradise.

The town smelled of heat and sweat, stale liquor and earth. It felt like a carnival was about to erupt at any moment.

Let me tickle your senses and make your brain dance. Palm trees hung in the shape of J's planted upside down, casting long, skinny shadows across white sandy beaches where locals drank cold beer or warm tea on hot days. You could swing from a vine into the deep blue ocean, shouting "Hakuna Matata," or make friends with beautiful Latin women with exquisite asses until night turned back into morning.

Or at least, you could try. And did I ever try.

My girlfriend and I had been fighting for the past year. It was all laughter then tears then laughter then tears while we fought. It was a bad teen soap opera, and good God, I was sick of it.

You know how seemingly everyone knows something is done before the two people living it? Why is that? Why can't you see clearly in the present? Why does love only seem to exist as perfect in the future or unreachable in the past, but never prescient in the present?

When I first arrived in Ecuador to work on my dream, I felt filled with purpose. And what's happiness without the ability to share it? So, I told her how I was. Waxed lyrically about how I was one step closer to paradise, to me and her slurping down Pina Coladas for breakfast like people too wealthy to ever be called alcoholics. I said I would buy a gun. Not just any gun, but a golden gun with skulls carved into the barrel. I would waltz around my half acre, a half drunk golden God. She exploded like a volcano, all burning vitriol and magma. She exploded because she wasn't there with me, enjoying the moment with me, and if she wasn't happy, then, well, why should I be?

They say absence makes the heart grow fonder,

but I don't know. I have a pretty good argument that it doesn't.

Women have their ways to carve out a heart. It's easy when they're given the tools to do so—a rusty spoon, I's, and Love's, and You's. Not every woman. Of course not.

Why do I always have to make that clear? If you're not capable of that, then good for you!

Why are you upset? This story isn't about you. It should be obvious that some women don't have the twinkle in their eye in the same way a cat does when it watches the fish move in its tank. But some do.

Just like some dogs bite, my girlfriend could carve out my beating heart and hold it there in her palm, free to poke, prod, squeeze, and manipulate until boredom set in and she stabs it and burns it with gasoline as her smiling face says, "Who? Me?"

Anyway.

After we built a concrete road from the town to my plot of land, I was tapped out of cash. I sat at a bar and drank a full day away as gorgeous strangers danced temptation around me. I couldn't join in, because I knew I would be sucked back in and never follow through on the hard parts of the dream.

Diamonds come from coal, put under extreme

stress and pressure. That's the part rarely sold in fairy tales. The part where it was time to go back, to afford my dream in instalment payments. There's only one place for beautiful bastards like me to find the kind of cash I needed.

As soon as I arrived back on Canadian soil—well, after custom agents checked my pockets, bags, and ass for drugs—I was free to go back to work. I knew I wouldn't piss hot, so I called the big cat: himself. Rossko.

Rossko was solid. Pick your definition of the word. He was a general manager of the company Little-bore, a three-hundred-and-fifty pound slab of beef with legs. His head was nearly a perfect circle, divided in the middle with a wonderful moustache, thick and dense. If you shared a beer with him, he would have your back forever. He was the foundation of the company, and he was able to retain hard workers, season after season, because he trusted us, and we repaid that trust in kind. Everyone was there because of him. No Rossko, no company, no Ecuador.

"Rossko, you cocksucker, I'm back. Back and better than ever. When are you gonna put my ass to work?"

"Here I am, about to enjoy a nice meal, when this

skinny-ass boy calls me and starts telling me about his ass," he says. "How was the trip? You pop your cherry?"

"You know me, Rossko; saving it for marriage, my friend."

"That's my boy! When you get over to the office, we can talk church before I ship your ass to work," he says.

The beach was both behind me and in front of me. I loved it in the past, and I'll love it again in the future, but the present is a grind. The bush. The pipeline. The madness of another shift in the patch beginning and ending with lunatics of all flavours. Another season. Maybe one more. Maybe one after that. Then I'm gone. Then I'm gone…

The Boys Are Back

The Big Rig is getting sent out. The tank. Rig 666. The 220 Vermeer, our baby. We need men. We need heart and soul. The comrades that will bleed for the rig.

We have no mothers or fathers anymore. We are the orphaned deadbeats of madness and insanity. But, just like Rome, we were all the riffraff, the outcasts, the lunatics who banded together, had each other's backs. That was our strength. Beer was our cement.

This was our rig family. We take care of our own.

I feel like I'm wading in grade-A gold. My favourite people in my oil and gas world are back together. Some of the best workers, best drinkers, and best men I know.

We've got a good core group of trustworthy bonds. That being said, we have some new faces. Some are children; some with lying, shifty eyes. Some with fragile, opiate egos. Some are pure acid casualties. Everyone has a purpose on this rig. Only a select few will ever make it on this rig, and if you won't die for this rig, then what are you doing in oil and gas?

Kindersley! That's where it will all begin.

I arrive on site in Kindersley, Saskatchewan, population under 5000. To tell you the truth, I'm pretty excited. The land of the flat plains is always better than the tundra. I don't like being up north in the land of only snow and oil. Everything expires. It's too finite. Driving the prairies is like driving on a treadmill. A man can see a thousand kilometres one direction and turn around to see the thousand kilometres he just drove. They say if you fall asleep while driving here, nothing will happen, because the land is flatter than piss on a plate. For days you could sleep at the wheel and just keep going on and on, not hitting anything but grain and dust. Maybe a house or a tractor would get in the way, but odds are still good that you'll wake up in a canola field without a scratch.

It doesn't matter if the population sign changes up and down depending on you entering and leaving. It doesn't matter if it's a town of twenty cousins related to the same mother with sisters and brothers married where the auntie is the mother to her sister's child. There is one thing about Saskatchewan that you can always bank on … Well, two things. First, you can always bank on seeing someone in a Roughriders jersey drinking at a local chain restaurant—Boston Pizza is a go-to. Second thing you can always count on are the town staples. Every town in Saskatchewan comes equipped with a Tim Hortons off the main road —usually the highway—a church, and a liquor store. Our church, a place of cold beers and 'hell ya's! Amen to that. To the liquor store, not the church. I've never been big on the church. I'll leave that to the family I mentioned above. But I have a certain affinity, a higher calling, if you will, to a store with ice cold beer. I intend to drink this small town dry.

The Chief

We're like Guinea pigs. We're social creatures. If kept without a group, a single guinea pig will often lonesome itself to death. I think humans are similar. Everyone needs a group, even if it's just to have some people to get away from for a bit. How do you blow off steam if there's nobody to return to? What's the normal you're escaping? So, even though I'm in Saskatchewan instead of Ecuador, it's good to see the boys.

Here with me are my boys: Booby Mac, Micky Makoy, The Chief, and my best bud Wilton.

The Chief is the head boss. He's the tool push, which means he's the guy running the show. He's a member of the Cree nation, festively plump, and when he walks, his ass jiggles like Jell-O. He has a crooked smile with a gap tooth right in the centre that stares you in the face when he talks.

"Cree Nation, my friend, what's going on?" I say.

He looks up from his phone and smiles warmly. "Well, look who's back. They run you out of Mexico already?"

"Ecuador."

"They ran you out of Ecuador, too?"

"How's the work?" I ask.

"You know how it is. Same shit, different day." He gestures out to the work site and the men milling about, finishing their morning coffees with a wide wave of his arm. "Look at you now, bud. You're darker than me. Should send you to the reserve; tell them you're my cousin."

"That doesn't sound so bad, does it?" I say. "So, what's the story here? What'd I come back to?"

"We're just setting up the 220. You'll be running tanks."

"Who we with?"

"Booby Mac, Micky, and P-Unit," he says. I can see him waiting on the next question.

"P-Unit?"

"Ah, yeah. God. He's my half-brother. Bit of a, uh, space cadet."

I wrinkle my face in annoyance. "Exactly what we need."

He puts me right to work setting up our drill. People probably don't know this, or maybe they do but never think about it, but when you try to put a pipeline in the ground, it doesn't slip right in like a hot knife through butter. Things get in the way all the damn time. And I don't just mean some trees and rocks. I mean, we'll need to get through roads, rivers, swamps, even lakes. Oil doesn't wait for nature. If it did, I wouldn't have a job.

If an excavator isn't enough for the job, the pipeline company brings us in. Today, we brought the Fuck-Off Drill. The big! The powerful! Weighing in at a whopping eighty-eight-thousand pounds … give it up for the 220 Vermeer Directional Drill!

Man, I love this thing. It's a tank. A beautiful, bright yellow tank with tracks, but where a tank would have a barrel, the Vermeer has a derrick with a big-ass top drive.

A derrick is what moves the drills up and down. There are two types of derricks. Type 1 is for drilling wells that go up and down. And there's the other type, Type 2, which is driven horizontal for going under creeks, rivers, and roads. I'm in the Type 2.

If you've spent even twenty minutes in Alberta, you've seen one. It's the framework for the pipe. With a top drive, it moves in a track along the derrick. Basically, the top drive are the big pieces of metal that move up and down the derrick that turn around and around and forces the pipe into the ground. We put the pipe into the derrick, top drive screws into the pipe, which then screws into a drill bit. The drill bit shoots out mud, which then comes from my tanks, and cuts and cleans the newly made hole. Then we place another pipe on the derrick and drill that pipe into the ground until we get to the other side of the creek or swamp or road. All of that comprises step one.

The next step is to put another big-ass piece of steel called the reamer where the drill head was. This one looks like a traffic cone covered in sharp teeth. It eats into the ground to widen the hole so we can pull the pipe through it. These pipes are connected on the exit side of the shot, which is the opposite side of the drill. I've seen a thousand different versions of this pipe, and they can be anywhere from one-hundred metres to three-thousand metres long. Just weld a bunch of small pipes together until you get to the desired length. Whatever distance the bore shot is, that's the length the pipe is going to be. The pipes then snake their way underground until we connect them to a bunch of other pipes and make a pipeline. That's how we bring all that fine gasoline, diesel, and crude oil to the refineries and to the gas station you like because it sells your preferred brand of smokes.

Now, there are many things a drill needs. Mud tanks and a machine that can recycle the mud—this is called a reclaimer, which is part of the tanks. That's where I come in. I'm the mud man. We also need a pump to take the mud from my tanks and send it down the hole to make sure it's lubed and keep the drill bit from getting too hot. This is called "cleaning the hole."

We need a tool crib, obviously, for tools. A generator for power, because you need a lot of power. A *lot* of fucking power. And the most important ingredient is a bunch of wonderful idiots who make all this shit run and fix things when shit breaks down. And after we put the pipes into the ground and the job is done, well, bing, bang, boom, pay us the money.

This is how you get the freedom to go from point A to point B. Isn't it fascinating? Think about that the next time you hop in your Camry late at night to drive with the windows down, free from the sound of crying children. Or when you road trip with your new girlfriend to visit her family and realize that she's the type to sip coffee with both hands and say *ahh* after each sip, but you're only halfway into the fourteen hour drive. Maybe now you'll stop asking why the price of gas is above a dollar.

Permits, Permits, Permits

Given the chance, someone would find a way to permit your exhale and charge you for each inhale. I'm stuck in permits. I'm drowning in permits. I have become permits, destroyer of worlds. I'm a soldier in a never-ending war between contracts, company, and government regulations.

We need a permit to move in our equipment.

We need a permit to bring in products.

We need a permit before work starts.

We need a permit when work ends.

We need a permit to smile.

We need a permit to fart.

We need a permit to laugh.

We need a permit to cry!

The rule makers and bean counters have taken all the fun out of work. We need to be robots! Machines! It'll drive a man mad it's so frustrating! Drive a man to drink! Drink a man to drive! If you break the rules, if you piss without a permit, cough without a permit, die without a permit. Graveyards full, it's illegal to pass on, so hold on a few more years, gramps! Common sense ain't that common, and now everyone has to pay for it. Why do we build our society for the safety of the lowest common denominator?

Banty. Banty is the guy who cursed us these permits. Another cousin from Cree nation. Boy, did he get lit up last night. I'm not saying drunk. Drunk is fine. Half of us are probably legally drunk each and every day. Banty, well, the man was fucked up. Just the way we like our pipeliners! But only if they show up to work the next day. Show up for work drunk and key bump yourself back to life for all we care, but you still need to be present at first call. But Banty was more dead than alive by morning light, so now our crew of ten are waiting for someone to come sign our permits. Waiting with our fingers up our asses. Bunch of idiots in a field, paid by the hour.

We're watching a permit battle take place between our consultant and The Chief. I'm up on the tanks, watching from afar. I can't hear anything; can only see their mouths moving and a spare hand gesture here and there.

Georgeo, our consultant, keeps bringing his hands to his chest then throwing them above his head, sort of like how Kermit the Frog runs. Oh! Georgeo throws a hard finger push into the chest of The Chief! And The Chief responds with a puffed chest! What action. I'm not an expert at lip reading or anything, but I'm pretty sure I get the gist of what they're saying:

Georgeo: *Yes, I keep the permits up my ass. You can reach up and grab one. But I have to ask ... do you have a permit to do that?*

The Chief: *I need a permit to reach up your ass and get a permit?*

Georgeo: *You can't work until you get another permit from out of my ass.*

The Chief: *Can't you just give us a permit? If you have so many stored up your ass?*

Georgeo: *Oh no, not at all. I like storing permits in my ass. Perhaps you can have one, but you'll need to reach up there yourself. Of course, you'll need a permit to do so.*

And so on.

I'm not sure, but I think that's what the conversation is about. Just a guess … just a very educated guess.

Booby and I have worked together for the last two and a half years at this job or that. He's the most knowledgeable guy I've come across in the industry, and he's got the go-getter spirit of a guy with one-fourth his know-how. It doesn't matter if we're baking like lizards in the hot sun or watching our breath freeze in front of us, he'll be bare-knuckled, elbow deep in the elements, fixing a breakdown with hydraulic oil pouring all over him. After working a fourteen hour day, he'll drive all night to grab the parts needed to fix a rig, turn right back around, and put in another full day without a wink of sleep, no complaint.

He's married to a wonderful gold digger. She's got a pretty face, long legs, full titties, porcelain skin, a body that comes from never working, never worrying. Booby's living the pipeliner dream! Big, fat house, jacked-up truck (Ford F-350 Platinum – only the best), 1000 Maverick Turbo side-by-side, and the fancy camper to take his gold digger wife into the elements in comfort once or twice a year. I admire the guy; he's one of my brothers.

More drama unfolds as Booby Mac and Georgeo get into it. Old Booby starts walking towards my tanks. Mudland, I call it. My home, and I'm fiercely territorial. But when he comes up, I know the look in his eye. He gets a serious twitch in his face when he's angry, like he put his hand on an electric fence. Right now, he looks like he has Tourette's. His eyes are going buck, and the side of his mouth is going bonkers.

So, I ask him with a big fat smile, "how the hell is it going today? Beautiful day, is it not? God blessed us with today."

"Do you know what that fucking retard wants me to do? Do you know?"

"I can't say I do."

"He wants me to write down every minute of every day. How am I supposed to get things done if I'm writing all day? He wants everything: when we drill, when we stop, when we drill."

"Do you have to write when you take a piss?" I say. "Is it a special form if you have to take a shit?"

"This is fucked," he says. "Me and him just had it out, man. I'm about to punch this fucker right in the throat."

"We've got some beer in the truck, you could try that instead of punching Georgeo," I say.

"Which truck?"

"The one that guy is leaning against," I try to make out who it is. "Booby, what's with that kid over there? Who is that?"

"Oh, you don't know? That's P-Unit. Useless. Only thing he can do is get you water, and you've got to specify you don't want it frozen. Useless doesn't begin to cover it. Don't even try; he'll only disappoint you. Trust me."

My First Year

Every year, thousands of little Albertan boys graduate to their first big boy Tonka truck with four-wheel drive and head up north to make it in the rigs. It's a rite of passage. And every year, about ninety percent of them come back home and tell stories in bars about how difficult it was, how isolating it was, how hard they worked for three months. Either you learn or you leave. Either you don't last the year or you're here for life. I went into the pipeline as a little boy and learned quickly that you either fit in or fuck off.

I was a little bitch. And that truth hurts. I had weak, spindly arms and skinny boy abs. Couldn't even drive stick. What can I say? I was a nice person. I really was … the perfect citizen. Weak, weak, weak. The wolves love the weak, the wolves feed on the weak, and I was a delicate lamb.

In my first year, everything I knew about the world and my place in it was changed. It was lonely. It was painful. It was beautiful. A brutal, cruel, chaotic world, built for the strong, built for the insane, built for wild animals. And I didn't get a word in edgewise for months.

"Hey, little bitch, get me a wrench!"

"Oh, are you tired, fag? Let's go. Stop your dog fucking."

"Are you cold? Oh man, why don't you go warm up like a bitch. Fuckin' hell. Where do they find these kids?"

"You silver spoon fed spoiled shit, you think you're better than me?"

"No such thing as a dumb question. Well, except that. You just asked the first dumb question I have ever heard. Get the fuck out of my sight. In all my years, I've never seen such a pathetic human."

"Clean up this mess, boy."

"Hey, dog fucker, get me a bucket of steam."

"What are you going to do? You wanna fight?"

I was so fucking sick of these people. I hated every single one of them. At times, I would dream ways to kill them and make it look like another industrial accident. I had forgotten the sound of my own voice. I had to change things or slip away into nothing more than a pale shadow that men cursed at when they stubbed their toe.

I started to work out. Went to the gym every day after work. Started to hit the punching bag, channeling every time I was called 'dog fucker' into a fist and trying to drive the bag clean off the chain. The nice, polite boy had to die, and I wasn't going to let these pricks do it. No, I would kill him myself. I would drown him in sweat or work him to death every shift. I would put the metaphorical gun to my head and pull the trigger. I was my own executioner.

I began seeing the world in a simple way: the weak and the strong. And I was sick of being weak.

"Hey, bitch, clean up that mess!"

"What'd you say, faggot?" I replied back, ready for the repercussions.

"What you say?" he said, sounding puzzled.

"I said, I can't hear what you're saying with all those double chins you have, you fat fuck."

The guy was two-hundred-and-forty pounds of fat. He would win the fight, I knew this. I also knew he would feel me in the morning.

My adrenaline was pumping! I was ready to die with a voice.

He gave me a head nod and called me by my name before adding, "clean that fucking mess."

P-Unit

A kid in a man's world can be a frustrating thing, for both the boy and the men. I've worked with P-Unit now on a few shifts. The Chief's half-brother, bit of a "space cadet" was the term he used. Goddamn. He's nothing but a fucking kid.

I don't hate P-Unit, but he does represent what is wrong with society in first world countries.

When you go to work, there's a job to be done. Can we agree on that? Every single guy down to a man has better places to be. We miss our family, friends, our lovers, our free time. Even if we have nothing in this world, we'd still rather be doing nothing right now than working. I can tell you a million other places I would rather be than here. I'd rather be in paradise, or a bar surrounded by naked women, or a bar surrounded by old, fat drunks, or looking at an exquisite painting done by Dali, or in a boat on a lake where all I see is water so far in the distance that it might as well meld with the sky for as much as I'll ever know about. Pick a place, any place, and I'd rather be there.

But P-Unit thinks that, out here, in the brutal cold nothingness of the prairies, he's the only one who doesn't like his job. It's almost sad. He's The Chief's half-brother, which means nothing to us, but he thinks it entitles him to some form of protection. Actually, it's worse for him. There's no love out here. There's respect, and it's earned, not given. If he was looking for a normal work environment where proximity to the boss meant a certain type of regard, then he came to the wrong place. I almost feel for the guy … almost. He's going to be broken and rebuilt just like I was.

There are laws of nature. The strong prey on the weak. The strong don't eat the strong; it's not worth the risk.

There's beauty in everyone out here. Even P-Unit. He had an innocence about him; a kind of impish charm. His shoulders slumped forward, and his skinny white arms and little wisps of red hair that stuck out from his chin made him look like Shaggy, the white Shaggy. Sometimes he makes me smile, like when he complains so much it turns in on itself and becomes funny, as though he was knowingly playing a character of someone annoying. But we work with hundreds of thousands of pounds of earth and mud and water and machine, and a boy slack-jawing into the sky, imagining his next sleepover to play video games with his friends, is a liability. If the boy inside him doesn't die, either by his own hand or from us, it'll be trouble. I can feel it in my bones. Black clouds are brewing. A storm is coming.

Micky Goes A Little Buck

We arrive at the Kindersley Inn Hotel—Micky, P-Unit, and myself—where we're lodging. We've taken to calling it the K.I. Sounds like it's from a spy movie. Makes it seem like it isn't the only hotel in town.

One beer, two beers, three beers. Back to normal. I feel the day melt away behind me and start massaging a knot in my left shoulder blade that has nagged at me for about two years now, on and off. Then I throw beer money on the table and leave to my room for a shower. Micky hits the lounge, and no one asks where P-Unit goes. No one cares.

All freshened up, cologne slapped on my cheeks like they do in the movies, I pop down to the lounge, finding the boys there—I can hear them before I enter the room. They're ordering food, telling war stories to guys from other crews who are there. Those guys are one-upping my boys who are two-upping them back. My very best friend, Wild Wilton, is there with an open spot and a cold beer for me. Fuck, what a beauty.

"What's going on, Wilton?" I ask as I sit beside him.

"Flanging hard, hard, hard, hard!" he says.

"How was the day?"

"Good, really good. The inspector is deadly out here. Oh yeah, he likes us," Wilton says. He radiates calm and good times. It's like sitting beside the human version of having a day off.

"Lucky you. The Chief went off on ours today."

"Yeah, well, he's probably a cunt. Who doesn't love The Chief? How was your day?"

Micky throws himself into the conversation, arms dangling over the table between Wilton and me. "It was a pretty, pretty good day," he says. "Except … P-Unit's kind of dumb. But pretty good."

"He's not that bad," I say. "Just a kid."

"He's a fucking retard," Booby chimes in.

"Wilton, he's not that bad, I swear. Just young," I tell him.

"Nah, for once, I agree with Booby. Pretty sure I saw him drool on himself today," Micky says, laughing. He goes to swig his beer, sees it's full, and acts offended by that. Then he chugs it down. "Excuse me, can I have another beer?" he says politely to the air.

"Make it a round!" Booby hollers in the direction of the server.

It's been about twenty minutes since I left Micky, and he's aged a half-dozen beers. He's pushing the beers back fast and hard.

"Hey, we didn't order these?" Micky smiles as shots of fireball accompany our beers to the table.

We got the boys from the pipeline crew at another table sending shots our way, which seems like a nice gesture, but it's actually a declaration of war on the pipelines. They started a battle of shots, and Micky figures himself a goddamn Napoleon.

It's a tricky skirmish to start. Both tables win for many rounds, but in the morning, we all lose. We quickly counter their fireball with tequila, but they fire back with JD. *Shock and awe.* We retaliate with a barrage of shots, hitting hardest with well vodka. The battle continues. The ones who don't want to face the battle-end of a toilet, pay their bills and toddle off to bed.

One thing about war is true: a great causality of war is freedom of expression. And I haven't spoken a coherent sentence in almost an hour.

I've loved before, and I've been proud to see my

friends love others—beautiful people with gorgeous traits—but no man or woman can compete with a drunk's love of the liquor, and the liquor is putting hickeys all over Micky. The beers and smokes have complete control.

The boys are going to the bar next door, and I stagger behind, a silent passenger to the mayhem. Shots are bought by the tray. They're ordering obscure shots in random amounts, counting into the air, "I'll have one, two, three … twelve shots, please," then drinking the difference. Wilton, Booby, and Micky are fighting with their wallets.

"You call that dancing?" Micky yells at Wilton, his voice rising above the music.

"No, *this* is dancing," Wilton says, and he starts elbowing the air and trying to limbo an imaginary bar.

He still fails.

"Catch me, catch me," Micky says. He backs up and sprints at Wilton, lunging at him like he's in *Dirty Dancing*.

Wilton shields himself, and Micky ricochets off him and lands star-fished on his back.

The floor is chaos. Two good ol' boys, one with a Packer jersey, who earlier corrected me angrily when I called it a Roughrider jersey, throws the drink of a guy wearing denim all the way down.

Never mess with a man wearing a Canadian tuxedo, out of pure respect.

He might have buried a keg that night. Suffice to say, Saskatchewan James Bond drops Brett Favre with a right to the jaw, and the Kindersley crowd erupts, their cheers only muffled by the grumbles that the fight wasn't longer. No one helps the poor guy off the floor. Everyone just goes back to their conversation, back to their beers. He lays still, the floor his bed and his pillow.

Sweet dreams, Gunslinger.

When the ugly lights come on and the bar starts to clear out, Micky is a complete write-off. He's happy drunk, stumbling with difficulty like his legs are a new car and he hasn't worked out the brakes. He takes to crawling: more effective.

In the bathroom of his hotel room his body hits the eject button on everything. His limbs fail, his stomach fails, and he hits every part of the small bathroom in one miraculous tumble. His forehead's bleeding, finger janked to the side, knee smashed. He curls up beside his vomit like it's the little spoon.

Sweet dreams, Micky. Work is in an hour.

Squeegeeing Grass, Shovelling Steam

People come and go; vacation, quit, or fired. You stop paying attention to the passage of people.

Have you ever seen the same stranger twice? You see a guy at the coffee shop and, a week later, he's behind you in line at the post office? It's like that. A look of recognition, a how-do-you-do nod.

Somebody left to go on vacation. Mexico, I think. Somewhere warm and sandy that made me yearn for Ecuador. Anyway, I had to leave my home on the tanks and cover for him, but I got to work beside Wilton.

It's pretty good working with Wild Wilton. He's always got a smile, quick with the perfect word to a waitress for a free round of shots. Six feet tall and constantly fretting about losing his hair.

I'm no expert, but I can't tell the difference between before and after he went to Hollywood for a seventeen-thousand dollar hair job. Couldn't tell the difference between the seventeen-thousand dollar hair job and when, a mere two years later, he got another one for four-thousand.

Back in the day, he was a hooker-loving machine who didn't like cocaine, just loved the smell of it. When you needed to borrow twenty dollars, he'd give you fifty and tell you to spend the extra on something that makes you feel pretty. Now he's got a wonderful girlfriend, and life is good for Wilton. Makes me damn happy to see him with such a fine, loyal woman.

He and I have been around the world, all the way to Aussie land and through the exotic plains of Asia. He helped me volunteer to teach English to Cambodian children adjacent to a rice field. Then he and I pedal-biked four-thousand kilometres across Asia, starting in Semi Reap, Cambodia. As weed smokers and beer drinkers, every day we suffered for the one-hundred-and-twenty kilometres we did. Ass pulverized by the seat as we made it to Saigon, then the brutal climb towards the Himalayas, crossing into the Chinese border, before heading farther east to Hong Kong. To celebrate this pointless accomplishment, we ordered tall beers and warm Saki, and boy, did we party as we climbed the stairway to heaven. He's a blood brother. I would kill for him. I would die for him. First one, then the other.

I have a secret for you. When you ask a man how he's doing and he spreads his arms and says, "Living the dream," I can guarantee he's already imagined his own death that day. And can you guess how Wilton responded when I asked him how his day was going? Arched way back and gestured to all his conquered nothing, he was living the dream.

"You know why?" he says. "Because I've got the barbecue going. I've got burgers and hotdogs. She's fired up hard, hard. Harrrrrd."

He's always cooking for the boys—burgers, steaks, hotdogs, pork, a slow cooker making chilli. If it was on four legs before dying, then Wilton is cooking it up for the boys.

"When's it going to be ready?"

"Hold your horses! Which I also have. Things take time. Can't just be cooking things half-ass around here. Quality control. There's an environmental inspector! Permits, my man. Takes time to do it right."

"When's it going to be ready? I'm hungry!"

The smell fills the air—barbecue and pipelines, dirtbags and scoundrels. Anyone with a sense of smell starts to circle the barbecue like a pack of vultures. Wilton is calling everyone over. "Come and get it.

Open season."

It felt right out of a picture book. But today, he was only cooking as a means to wait out the inevitable damage he knew was coming, and after the feed, it's back to work.

There's pipe above the ground, which is ridiculous. What's it doing here? It needs to be below ground.

We're on farmland. Soft, airy, flimsy farmland. It's great for growing crops, but it's shit for the type of drilling we're doing. Old timers tell me, back in the day, they wouldn't have dealt with this type of land. Would have avoided the whole hassle and found a workaround. But times have changed, and the people who know the industry- the guys with twenty years under their belt- have no say. The word of gospel comes from an engineer at a desk in Calgary, feeling a little thrill because he's wearing running shoes when the office says business casual attire, who tells us how to do our job. This turns a job that could be done in four days into one that takes two weeks.

This industry's too rich, but they cut corners in the wrong ways. If they can see you, in the office, at the parties, on the golf course, you'll survive. But they'll cut money for us out here, dying hourly.

Anyway, the all-knowing engineers in Calgary want us to use high pressure and lots of volume in a light aired soil. Think about it. Think about it even a little. Do you see the problem? Think of blowing up a massive balloon below a thin layer of debris. Unless you go deep into the hard ground, all this fluid goes up … up … up … up to the surface and explodes all over. The more you pump, the bigger the mess. And oh, sweet baby Jesus, did we make a mess. Beautiful, thick drilling mud, just fucking coating this farmer's field. Everything is shut down instantly, complete standstill. And now we have to clean up the mess.

"Imagine that fucker at his stupid desk with his starched fucking shirt," Wilton bellows. "Just look at this shit! What the fuck was he thinking?"

I yell at P-Unit, "Get the frack trailer. Grab sawdust bags, get the Hydrovac … And some squeegees, goddamn."

"If I ever meet this engineer, I'm gonna do some damage," Wilton says. "Why the fuck are we pumping so high in soil like this? Did he not believe us when we said this would fail?"

"He thinks we're dipshits, man. Dumb fucking rig pigs."

I love my job! I love my job! I love my job! I

repeat this mantra for the crew as we all kneel like assholes with squeegees in our hands, squeegeeing a farmer's field. A useless farmer's field in goddamned Kindersley, Saskatchewan. Fuck my life.

We look like a bunch of morons, squeegeeing mud on top of mud, pushing it limply to the Hydrovac's hose. I should have stayed in school. I should have become a painter, a sculptor, a fucking banker, worked in a goddamned 7-11, for fuck's sake.

The difference between drilling mud and the farmer's soil is that drilling mud comes from a place called Wyoming, USA. It's called bentonite. It's a different type of mud, a separate breed of soil. The funny thing is that, after we're done, the farmers ask us to spread this American mud on their fields, because it fertilizes the crops better. So here we are, squeegeeing this USA mud on top of Canadian mud. The environmentalist, sitting in his office in Calgary, who sold out and defends his choices to girls with dreads in cafes, wants us to keep on squeegeeing the dry soil over and over and over again. And isn't that life? Spreading shit on shit for someone, somewhere else, who I'll never fucking meet.

A Fallen Soldier

Anything can kill ya. You can cross the road and *BOOM*, get rammed by a Volkswagen Beetle full of sixteen clowns. You survive the impact, but then all those tutu-wearing, horn-honking, red-rubber-nosed jokers storm out one by one, and stomp on you with their big red shoes. It happened to a friend of a friend of my uncle's, hand to God.

My point is that you shouldn't focus your energy on death. It can happen anytime, anywhere, anyhow. Most people are scared of everything, from swarms of killer bees colonizing their way up the coast to herds of wild pigs storming into your home and munching your bones, but most of the time, we are scared, and nothing happens. We clock into our jobs, 'rebel' by pooping on company time, then clock out to go home and eat Swanson's in front of the television while we worry about what could potentially kill us.

Back to reality. We were drilling away, laying pipe in the ground by day and drinking beer by night. Standard fare for a few weeks. Then, one night, we were told to head home early from our shift and rest easy.

In the morning, all of us gathered around and stood in a half circle in front of someone we were told was an outside consultant.

"Gather close, boys," he says, waving us closer with an air of seriousness. "Listen; this is not your normal chat." His tongue pokes around his mouth, trying to dislodge the proper words. "There was an accident yesterday and we just need everyone to know so we can make sure it doesn't happen again."

A couple of guys in the back crack a joke under their breath, and a few loose giggles travel to the front. The consultant chooses to silence it by sharply spitting out his next sentence. "We had someone caught behind a truck as it was backing up. He's in critical condition."

The laughter stops, but now I can see the gears turning behind everyone's eyes. The job of the guy behind the truck is to make hand signals and guide the driver. How the fuck do you get run over by a truck backing up? It's going, like, five. Just move to the side.

The consultant continues, "so what happened is the driver somehow got the floor mat stuck on the pedal. The truck took off, driver panicked, truck hit a frozen clump of dirt, the back jumped and came atop the guy. Crushed his chest. Stars helicopter airlifted him to the nearest hospital, but … well … they don't know. They doubt it."

Everyone stands silent, thinking. Like … what are the chances of a floor mat getting stuck at the exact same time someone is backing up and then hitting a mud clump? Who could predict that series of events?

I think about all the close calls I've had in my life; another inch here, a foot there, and I'd be smashed, minced, or smooshed… but this? This doesn't make sense.

The guy's name was John. John died the next day. He had two little girls.

A circle of pipeliners raised a beer to him, but somewhere, two little girls now speak of Dad in past tense, all because of a mat and a chunk of frozen dirt.

There's nothing to learn from this. This may have never happened before and may never again, except it happened to John, here, today. Sometimes, life makes you truly feel small.

Booze Cruises

All day, I'm up on my tanks. It's pretty easy when everything is going well. When everything's easy and everything's good, my mind starts to wander. I start thinking of her. Daydreaming … Daydreaming is dangerous. Daydreaming all the good and open, loving moments. The kind of moments that memory freezes in time to be lived in forever. Like when I got up early for work, and she woke up just to give me a kiss goodbye. How she fit so well in my arms at night, where I could breathe in her intoxicating smell. The look in her eyes that said I was the only one she wanted, that told me we were the only two things in the world that God didn't have His eyes on.

It hurts when my mind wanders. It always wanders to positive places, always in the past. We haven't talked for some time; two months and a day, but who's counting? My thoughts always sour, turn bleaker, until all I see is her crying, and it breaks my heart. I cry inside, but I have to hide it, only for it to come out when I'm alone in a dark place, shut off from the world, where no one can see.

Thank God for Micky. He breaks up my daydreams and snaps me back to normal. We can only communicate with hand signals and lip reading over the screaming of the diesel motors, and I see him clapping for my attention in my peripheral vision. His hand is to his mouth, in the international sign of "let's have a beer."

I shake my head *no*.

He looks confused, his bushy beard sagging with his frown. He puts up one finger. *One beer.* It's only one. El Corrupter.

I nod to the one beer and hold up one finger forcefully.

His damn smile widens, warm and inviting. It radiates the feeling that nothing bad is going to happen. Then, somehow, where there was once one finger, there's now two. Two beers.

I shake my head.

The finger goes back down, and his shoulders slump. His body looks like a balloon deflating. The smile disappears, and I feel empty. I want it back. It was for me.

I raise my hand and signal two fingers.

Micky practically bounces back to life—three fingers. I guess it's a six-pack tonight. What's the worst that can happen?

We finish the day after fourteen hours of working. *Fucking yeeeehoooo, quitting time, baby.*

We're having a couple of wobbly pops. I'm in the passenger seat, and Micky is in the back middle, poking his body through the seats so that he's a part of our world at the front.

"P-Unit, the truck looks great," I say.

"Yeah, I cleaned the windows, I wiped down the seats, I got rid of the garbage, then I ate my lunch, then I put all that garbage in the garbage bag, wiped down the dash, then the steering wheel—"

"For fuck's sake, P-Unit," I stop him. "I don't want your life story. I just wanted to say good job." I look back at Micky, who is shaking his head.

Now we are on the road, heading to the town of Durango. We have never been there, but like I said before, any small town in Saskatchewan has a church and a liquor store.

As soon as we roll in, truck tires kicking dust behind us along the road, it appears before our eyes—the holy grail. The most welcoming sign out there in all the world. "*We have cold beer.*"

"I'll get beer today, Micky. What are you drinking?"

"Surprise me. Bud or Bud Light."

"Can I have some?" P-Unit asks.

"You're driving! You can't have any," Micky answers.

"Get that country music ready, bud," I say. "This needs to be perfect."

I walk into the bar/liquor store, and it's exactly what I want it to be—dark and dreary, aged wood, lit by its own character. This wood has seen bar fights, blood, love; friendships have been made here, friendships have ended here.

The bartender has as much character as the wood itself. Face grained with wrinkles; some from smoking, others from age, a few from laughter. Behind her are dozens of neon lights, advertising long-gone years and events from different countries. "*Las Vegas welcomes you!*" says one in glowing purple.

I pick up some Bud and clamber back into the truck.

Now listen, I'm not a smoker. To any kids reading this: it's a disgusting habit. Waste of money. The smell it leaves behind is terrible. It's barely even a drug; it doesn't get you high. Slowly kills you, makes your teeth look like shit, ashes your mouth until everything you eat tastes like shit. Trust me; it's the worst … *except* … when you're drinking. God, drinking and smoking go together like rum and Coke, peanut butter and jelly.

So, here we are, Micky, and P-Unit and me, going back to the hotel.

Micky gives me a smoke, and I give him a beer, while country music plays in the background. Life is good. Hell, it's better than good.

"Micky," I say, not asking a question. "I've told you once, and I'll tell you a thousand times, you just can't beat country roads, drinking beer, while driving and smoking some cancer."

"Can't beat it. Wouldn't try," he says.

I ask P-Unit, "You want a beer, bud?"

"I can't. I'm driving."

"You fucking loser," Micky says.

"Okay, I'll have one," he says.

"You're driving, P-Unit. What the fuck, man?" I tease.

There's an art to a booze cruise. Gearing down from the day with some great guys, relaxing from the hard work as the plains roll by while country music soothes your soul and ice-cold beer reminds you why life is sometimes worth living.

We sing and smoke, drink and nod our heads.

I look back at Micky until he says, "stop looking at me!" with a chuckle and a wave. He absentmindedly picks at the edge of the beer label while swirling the last mouthful around the bottom. He pours it into his mouth as he rolls down the window.

"There's only one problem with driving a crew truck," Micky says. He leans his torso out the window, arches back and slings the empty beer bottle away from the moving truck, aiming for a sign warning of potential deer crossing. "There can't be any evidence!"

I chug mine down and then we both hang out the truck, waiting for the next sign, feeling the wind in our hair—well, not Micky, he has none—smelling the freshness of the prairies, holding the bottled-in excitement until I release …

The bottle flies gracefully through the air and smashes into a stop sign, shattering in the high grass with thousands of pieces of brown glass. What a feeling of satisfaction.

"Stop sign, 50 points!" I yell.

The truth is that we usually miss. Honestly, I think that was like the fifth I've ever hit. But it doesn't matter. Missing just gives the local bottle pickers something to do. It's pipeline charity at ten cents a donation. You're fucking welcome.

Micky's got that look in his eyes. Ah, shit. The beer is the key that unlocks him. But I can't say no to the guy.

I hand him another one.

We're almost at the hotel, the pack of beer is done, but I have a feeling Micky's just getting started.

Thanksgiving with the Rig Family

There are only a few days that are untouchable. Christmas, a day meant for family and friends, spreads of food, gifts, love, and Jesus I guess. Easter and Thanksgiving are meant for the same. That's all we ask out here, a handful of days a year to see our loved ones and have our kids remember our voices. Those sons of bitches promised us, and then they wiped their ass with the promise.

They told us, point blank, "you'll be home for Thanksgiving." Yup. Promised I'd be home to eat Mama's home cooking: mashed potatoes, roasted carrots, cabbage rolls, gravy—thick, beautiful, delicious gravy—and turkey I drown in that gravy. Not to mention the cranberries, scalloped potatoes, even the salad hits different on Thanksgiving. Then the drinks … Caesars to start, some cognac, Johnny Platinum, all with a beer in hand to boot.

But it's the family and friends I'm upset about missing. Drunk Grampa, drunk Auntie, drunk cousins… Family is better with a few drinks to lighten the mood. That brings the laughter, which brings the hugs, which brings more drinking, more turkey. Suddenly drunk Grampa is howling like a wild dog and things are starting to get crazy. Mom brings out the cake, then the pumpkin pie. *There's more? I'm stuffed. Apple pie with vanilla ice cream? You're too good to me. I'm so thankful. Hennessy? Don't mind if I sneak a little. After all, more drinks is more happiness.* Doesn't that sound nice?

But those boner-biting sons of bitches, shit-gobbling lowlife worms took a great moment from the boys. This is a big *Fuck You* to us, courtesy of a bunch of used handkerchiefs in a teenager's trashcan, donkey fucking, bladder infection shit heels. Making us work on Thanksgiving … I hate them! I'm going to drive to the house of whoever made the call for us to work while he sits in his favourite chair, warm in his favourite sweater. All his family there, enjoying their dinner. He'll be carving the turkey with his wife and children flanking him in their fucking dress clothes. Grampa Jebediah doesn't talk much, but even he's thankful they can all be together on this most special of days.

That's when I burst through the window like Chuck Norris. But I'll do it right, do it personal, not with nunchucks. I'll have a sixty-inch pipe wrench from the site, and I'll start beating the jizz stain to death right in front of his whole family. Little bits and pieces of him will be flying everywhere—a finger, an ear … Look at that, an eyeball fell right into his son's glass! I'm carving up the bastard like he's the main course, sending bits of matted hair all over his wife's cooking. They're all screaming, watching, except him: he'll be dying. Their Thanksgiving will be ruined, just like mine. I'll take both turkey legs and the wish bone on my way out, back to work, just like he ordered.

FUCK YOU, OIL AND GAS!

Cat's Cradle

"Hey, Mom, how are you?"

"I'm good! How are you, honey?"

I've never called and had my mom sound upset. I wonder if she always perks up when she sees it's me calling or if she plays it up sometimes, just so I don't worry.

"Fine … Well, actually, a bit shit. I'll be missing Thanksgiving. Bastards wants us to work."

"Those little shits!" she says, instantly furious on my behalf. Did I mention I love my Mom? "Is there any way to get out of it?"

"I could walk off site, let down all the boys," I say.

"You can't do that," she tells me. "But you know how important Thanksgiving is to me." I can hear the sadness in her voice. She doesn't get to see me or my brother very often.

"I know," I say. "But soon I'll be living the good life. Hot weather and sandy beaches."

"Well, did you ever think of spending more time

here in Canada? Ecuador just doesn't seem stable. Maybe you should just wait. Before you do anything, wait for the right time."

We have had this same conversation a hundred times. What my Mom and Dad want is simple: get a job in town, find a girlfriend, get married, fuck, and make babies. That's want they want. Then they will find something else to complain about.

Mom continues over the silence. "listen … do you ever think about getting a normal job? Not in oil and gas? Maybe then it would feel more like home, because it would be."

"I've tried that, Mom. There's no adventure there! It's like a retirement home, but instead of pills, we put back beers."

"Honey, I just don't think Ecuador is the right direction. Why don't you find a nice girl and give me some grandkids, damnit?" she says with a half-serious laugh. "Do you still talk to Nami?"

"Haven't in a while," I say.

"I liked her. You should talk to her again." Her soft criticism is crushing. Parents know exactly how to wound. "Oh! Here's your Father. He wants to talk to you."

Shit, I think to myself. I love my Dad but, right

now, he's just another parent unhappy with my direction in life. It's two against one. It's unfair.

"How's the pipeline world, son?" he asks in greeting.

"Good," I say, waiting for more soft disapproval.

"Long hours?"

"Yes, Dad. Twelve to fourteen."

"How's the weather?"

"Half-shit but not so bad."

"You ever think about getting a trade, working in a nice, warm shop?" he says with a chuckle.

"No, doesn't pay well enough. Need money for Ecuador, Dad. You know that," I say with the same forced laugh.

"Right, right … So, when you get your little thing all built, what are you going to do? Just … lay on a beach?"

"That's the plan. Maybe become a gangster, run drugs, guns, women. Whatever pays."

He murmurs, "do you really think that's going to be a fulfilling life? You're a half-decent welder. Why don't you get an apprenticeship?"

"Rather lay on a beach, Dad. Sounds much better than working in a shop," I say. "Besides, I have two properties in town so, hopefully, I can live off the rent, y'know?"

"You see, that right there, those are good investments. You should do more of that. It sounds … Son, it sounds like you don't know where you're going. Think about how much easier it would be to live in Canada."

"Fuck, Dad. I work half a year, then I head down south. I'm basically a snowbird already."

"I think the life you're living could be better."

"Yeah, yeah… Happy Thanksgiving. Sorry I won't make it home." I end the call.

I didn't enjoy the conversation.

Their words cut like a knife. Always about my direction and how it's never going the right way. Not many people can get under my skin. Some people can piss me off, but it's always temporary, always fades as quickly as it came. But my parents make me question every choice, talk me out of decisions that I've been set on for years.

I stare at the ceiling and space out on stucco while I ponder how my Mother and Father, my creators, know every chink in my armour and every tactic to exploit its weaknesses.

I want them to be proud and for me to be a son they can admire. I want to be spoken about in confident voices at suburban garden parties. I don't want to see their eyes downcast when I talk about another season in the patch. It's the only reason I have any property here instead of everything in Ecuador. I have nothing nice in my life, just investments of soil, cement, wood and nails, all banged, mashed, and mingled into places humans can sleep. I thought that would be enough to make them proud or, at the very least, shut them up, but I'm not sure if anything will ever be good enough.

Some people, they exist to seek more. It's like there's this maw inside them that is always empty, and they have children to fill it up, but it only widens the emptiness. It's not their fault. They always need more, always. Faster internet, newer cars, bigger beer cans, stronger coffee, hotter wives, richer husbands, living in more extravagant houses; never good enough, never satisfied, their thirst never quenched.

There's only one thing that keeps coming to my mind, over and over again. It's never stopped during the many years, not fully. It's what still drives me today, through all the bullshit, miserable days stretching into weeks. It's a focus on my purpose and my dream. Fuck all the distractions, stay on task, focus, grind your teeth, dig your heels, clench your butt cheeks, move forward, towards the dream, towards Ecuador. *You know what's best for you! You know what's best for you! You know what's best for you!*

Kentucky Bluegrass

Kentucky bluegrass is the best, most comfortable grass out there for laying your head down and watching the shapes of clouds as they roll across the sky like a movie screen. *Oh look! There's a dragon. And that one is a hand flipping the bird.*

Kentucky bluegrass is special. It caresses you, says, "Don't worry about a thing; you're here now".

In suburbia, Joe Blow or Rick the neighbour, who never seems to work, are always so dedicated to maintaining their grass. They'll say with complete earnest, "why would anyone do drugs when they could just mow a lawn?" I think it's a white guy competition, so they each spend hundreds on weed killer, soft fertilizer, and seed, and they put so much love into the lawn that every blade reaches up to thank them. It's as if it was made for sleeping on.

Goose Fest has been the talk of the radio and chain bars for weeks. It's October and the geese can feel the change coming. All these geese mingling around, quicker to snap than usual, because they know it's about to get really cold, really fast, and they're ready to get the fuck out of Canada and start their great migration to somewhere warmer and better. Give it a couple more seasons here, and I'll be leaving along with the geese to a warmer climate.

In a town where not much happens, Goose Fest is a lightning rod. It's set to take place in the agricultural dome in the centre of town. The locals are excited that the town council has booked some cover band from British Columbia called ACBC that only plays ACDC songs. People are moving through the week with a spring in their step, and even the pipeline boys are saying, "two more days, two more sleeps," as we work our way to Saturday night.

Of course, there's turmoil in Booby Mac's life once Saturday comes and we're ready to drink the night away. His old lady has upped her resentment, really turned on the hate, and has stopped messaging or receiving calls from him. I don't blame her. In the last year he's been working eleven months on site and been home for one. He says it's for her, for their future, but I think he's comfortable here, or at least more so than there. Work is simple, even when it's difficult. Life is complicated, always.

Booby Mac is a dirty combination of very

unhappy and incredibly torqued up. His voice is loud, his face wild with bad thoughts, a mean twitch in his eye.

We're all in Wilton's hotel room, spread out on chairs and double beds, surrounded by cases upon cases of Budweiser. All the boys are present—Micky Makoy, Booby, The Chief, Wilton. We got enough beer to supply an army, and there's a pile of cocaine on the bedside table that's gradually being siphoned away as the boys take turns doing lines with a red bill. I've never been a cocaine guy—I'm already too wired as it is—but pipeliners have hungry noses. Every crew has a few guys who are just fucking Dyson's, capable of hoovering up a few hundred dollars if given the time and space.

Micky Makoy isn't drinking, and it's giving the room a strange energy. It's especially puzzling to the boys hitting snowbanks. Cocaine demands answers, needs to fix what it sees as a problem, and Micky not drinking is a problem. He knows something we don't. He must know that the moon is full, that every two-thousand years all the planets align and demand a sacrifice.

"Mick, what the fuck, man? Why are you not drinking?" Wilton asks, rubbing his nose on his sleeve.

"Because, when I drink, I go crazy."

"But you have to have one," he pleads.

"If I start with one, I don't stop. They're just too good."

"I'll make you stop," The Chief says. "I'll make sure you behave tonight."

"I don't know … You know beer does things to people. You say that now, but later on, you're enjoying the Micky show."

Then a beer was driven into his hand. No one seemed to catch who did it. We all did it, I guess.

"Micky, just have one," I say. "What's the worst that can happen?"

"Christ, fine. If it's freaking you out so much, I'll just have the one," he says.

"Just one, Micky."

Micky cracks open the beer and the satisfying release of carbonation is met with a cheer from the whole room. Things now feel in their place. Micky not drinking is like a nun not worshiping the Lord. It's not right.

He takes his first step, and a dangerous, inviting smile creeps onto his face. I know what that means—he needs more, and nothing can stop him now.

The beers are going down easy and the pile of cocaine is getting smaller. The night is almost ripe, ready for the picking.

It doesn't take Micky long to catch up with the rest of us in drinks. It's about ten o'clock, the boys are feeling good, and Micky … forget about it. He is unstoppable, unmovable, a force to be reckoned with. Like the Greek God of the party, Dionysus, he's ready for a festival, a bacchanal.

"Hey, Micky, isn't this better? Beers with the boys?" The Chief says as we walk to the Argo dome.

"Yeah, they are pretty good. Don't know why I try to deny myself."

I jump on Wilton's back and scream, "Yee-haw."

He takes off bucking and running in circles.

"Hey, Booby, what's taking so long?" Micky yells behind us.

"The old jimmy leg is acting up," he says. Booby Mac has a bad sciatic nerve and thinks we care.

"Oh no, not the jimmy leg."

"If you would drink fast, that won't happen," Micky yells.

"Yer an idiot," he says, punching at his leg. "That has nothing to do with it."

"Did you get that from an old injury when they

converted you to a bitch?" I say.

"Or is it from when you busted your hymen?" Wilton adds.

"Shut up, shut up, shut up," Booby yells back.

The beer and the walk are soothing our souls, even Booby seems to be mellowing out. We are hugging each other, tackling one another, laughing as beer gets spilled on the sidewalk. We can hear the Goose Fest music, and we begin to feel the energy of the party.

Closer to the gate, we see some of the boys from the pipeline crews. They are right on their way to half cut.

Through the entrance, there's everything you would expect in small town, Saskatchewan. Some of the men are wearing nice clothes to disguise pot bellies. You want that super-sized? Hell yes, he does! They sweat when they dance and the pit stains and back sweat darkens their clothes, foreheads just leaking. And the woman … some look really good, but they are few and far between. The days of the tiny waists and double bubble are either over or never arrived in Kindersley. I see the hard work of Colonel Sanders in the rolls over top of other rolls visible through the crowd. Some have two rolls, others have six, all sweating in the lights of the stage.

The thing is, when you're working all the time, surrounded by a dozen dirty, sweaty guys, these women don't look too bad. They'll be dancing with one of those guys with big guts, and his gut will be grinding on her rolls, and I'll be getting another drink, watching Ol' McDonald dance with Wendy.

There are a few good-looking ladies, but that might be the beer talking. I get a dance in with one of them. Wild Wilton gets his grind on with a good old country girl. He's doing the classic chicken dance. There's no dancing quite like dirty dancing. Booby is putting back the rye and Coke's passionately, and that seems to loosen him up. There is no Micky! Where is Micky?

Wandering around Goose Fest is a trip. Everyone looks like an uglier version of someone I know. I feel like I've seen Micky a dozen times and right when I think I spot him by the outhouses, a girl starts making eyes at me from the edge of the dance floor. She has long brown hair, a cowboy hat branded with Corona, and passion in all the right places.

"Well, hello. What's your name?" I say.

"Melissa," she says. "Yours?"

"Batman," I reply.

"Not Bruce Wayne?"

"Not tonight."

"You dance pretty good, Batman. You from around here?"

"I sure hope not," I say.

She laughs and shrugs as if to say *it's not so bad.*

I'm compelled to find Micky, because I feel like he's going to bring down the house, but I'm horny. So, I try the direct approach. It either works or it doesn't; there's no real in-between. When you're direct, it shows you know what you want, and there's nothing hotter than knowing what you want.

"Melissa, I think you're absolutely beautiful."

"Thank you," she says.

"And you're turning me on."

"Oh, I am?"

I step closer and speak in a whisper, "Melissa, what I want to do, somewhere in the vicinity of that stage, is slowly pull down those tight black pants of yours, take my hard cock, and fuck your tight, wet pussy."

She just stares at me. For a while. She doesn't immediately reject it, though, and I can see the proposition seducing her slowly while she weighs different factors.

I look for Micky from over her shoulder while she sizes me up.

"Do you talk to all girls like that?" she says.

"Only the lucky ones."

"That's a bit forward, don't you think?"

"And?"

She lets me take her hand. I pull her in, and we start kissing.

Now, I'm no psychologist, but I think women are just as dirty as men. Some take it as sacrilege, like it can't be true, but find a girl who is hot and horny, and she'll be ready for some action that will make any pervert blush.

After some PG-rated groping, I get Melissa's number, promise to call, and then leave to find the boys.

Maybe it's true that the stars and moon and planets have aligned and a sacrifice has to be made—our boy, Micky Makoy.

The one beer turned quickly to six, which released any walls or thought. High octane rocket fuel was carrying him on a beautiful mission of self-destruction. Micky was gone, replaced by an angry sixteen-year-old boy who had raided his dad's liquor cabinet and replaced the vodka with tap water because he believes that, if he drinks enough, all his dreams will come true. He has to prove he is the best, the maddest, the craziest of us all, the biggest tornado to ever hit Kindersley, Saskatchewan.

I would catch him out of the corner of my eye, always with both palms laying on the bar, wearing a dangerous smile below shark eyes. He was on a level few have reached—Jesus, Buddha, my Grampa that one Thanksgiving. True enlightenment.

The last time I saw him, it was great. We were in the corridor and ran into the highest of all bosses, the hire and the fire, the big wig himself.

He tells Micky, "you better be at work tomorrow," with a half-smirk.

"Don't worry about me, motherfucker; I'll be there. I never miss a day!" Micky says, somehow slurring a laugh.

"Okay, you crazy fuck, I'll believe it when I see."

"I'll be there, cocksucker. *You* better be there tomorrow!" Micky yells. "I always make it to work, always. Do you, huh? Do you?"

We grab Micky and put our hands over his mouth, telling him to shut up through our own laughter.

When we leave, Micky tells us he's going to stay

a bit longer, he still needs more time to destroy himself. We give our salute and wave goodbye to Micky, who is dancing to a beat in his head, singing a song different than the one played by the cover band.

The rest of Micky's night has been recounted to us by other pipeliners and strangers at bars who point him out and say things like, "Glad to see you survived!"

Micky continued to drink like a man poisoned, possessed by Satan himself. Another beer was driven down his throat. Another rum and Coke. Then a random half-drunk drink from some table. It all went in … More, there's still more room for the liquor. More room for the madness.

It's time to leave. He thinks, *Which way to go? Which way to get home?* Life is hard now. His feet don't move the way they should. Time starts to slow down. He starts off on his long, hard journey home. It takes twice as long as normal, because he takes one step forward and two steps back, plus three steps to the side. When they find him, it's clear he was going in the wrong direction.

He's weighed down by the liquor. The burden is too great. His feet are cement blocks. It's time for a quick rest, just a little snooze for energy. Then he sees the perfect place to rest for a while: the sweet, inviting Kentucky bluegrass of Joe Blow's front lawn. He dives headlong into the grass and, just as promised, the KBG is cool on his face, the midnight dew like a shower for his worries. Everything is okay. Everything is okay.

There are people walking home from Goose Fest who walk by Micky and point and laugh. No one helps him, they just walk on by saying, "what a crazy dickhead." Eventually, the cops come.

I have to say, I'm not a big fan of pigs, but I'll give these dirty, filthy pigs a big Texas-sized 10-4.

Micky's on the KBG and can't move. The cop asks Micky to get up, but Micky tells them just where to go and "fuck off." Good job, Micky.

So, now both cops come and grab Micky and, as they drag him to his feet, he throws up all over this stranger's lawn, startling the cops who drop him right into his sick.

"Fuck you … That hurt …" Micky says.

Maybe the cops feel for Micky, maybe they've been there themselves, maybe they just hate paperwork, but whatever their reasoning, instead of bringing him to the drunk tank and slapping him with a ticket and a day of missed work, they take him to the hotel, right up to his room. That shows character. Makes me nostalgic for a time I never experienced, when people realized that sometimes a guy just needs to break his own chains for a while. Doesn't mean he's a bad guy, just means he had a bad night.

Sleep now, you precious lunatic; you'll need your strength.

Kid in a Man's World

We, as a society, don't insult children enough. They're dirty, filthy sacks of germs and lies. You can't trust a child, especially a stupid one. You can't trust them because they're half-formed, their brains not all there yet. They're weak and selfish. All of that is P-Unit to a goddamn tee.

P-Unit is a kid in a man's world.

In the world of the pipeline or the military or anything hard and dangerous, you need trust above all. If you can't trust the guy beside you, he might as well not be there. Trust is everything out here. It's how our connections work, our bonds of friendship are created.

When Booby or Micky or The Chief ask me to do something, they know it will be done complete one hundred percent. When I ask them for something, I expect the same, and I receive the same quality, every time.

This trust makes tight bonds, same as when you're in the trenches of war. You're not fighting for your country or money or fame, but for the guy beside you. You're a brother-in-arms, a comrade.

P-Unit isn't a comrade, and he isn't a brother. Every time we put trust in this boy, he fails us.

The worst thing about him is that he's tricky, frustratingly sneaky with his shittiness. He'll do a great job for two, even three days. Everything we asked, he will complete correctly and efficiently. Then it's like he gets complacent and the child returns, or maybe he disappears to smoke crack and it scrambles his brain for the next day.

I don't trust P-Unit. You can't be one of the boys without trust, and without trust, you're an outcast. When it's cold, you're outside, all alone.

At first, I didn't mind young P-Unit. I liked his innocence. He had a boyish charm that had been missing from the pipeline world. I thought he worked pretty hard, and I didn't really understand why the other guys were so hard on him. I took it upon myself to give the kid some confidence, raise his spirits so he isn't crushed until he's packing his stuff to head back home.

The first rule of pipeline world is that you can't be a pussy and a pipeliner. But the second rule, equally as important, is to put your fucking tools away.

When we have a disaster and need to stop nasty leaks or repair a piece of broken equipment quickly, it's important that our tools are all there. And P-Unit leaves our beautiful tools strewn all over the ground. I would be walking around and see a hammer over by the pump house, a socket set in the mud, pliers on the tracks of the rigs. BLASPHEMY!

"Hey, P-Unit, you're doing a pretty good job!" I would say with love in my eyes.

His face would light up. Any compliment seems like an oasis when you're in a desert of insults, a port in a storm of shit.

"Thanks, bud! Yeah, I've been working really hard. I've been putting garbage away and the truck is clean."

I nod and smile. He smiles back. I have him right where I want him.

My face turns dark, my smile morphing into a ruthless glare. "You pathetic fucking shit … who do you think I am, your goddamn mother? Am I supposed to be picking up after you? All I see is all our beautiful tools laying on the fucking ground. What kind of shit-show you running here?"

The smile runs far, far away from his face, and

his voice sounds childish. "What do you mean? I put my tools away."

I know what's in his head. It's almost not his fault. It's full of what his mother drove into him as a child—it's okay to do half-ass work, it's okay as long as you feel okay. Everyone gets a gold star.

"Let's go for a walk," I say, gesturing him to come with me. Then I throw my arm around him and squeeze him tightly before pointing out a screwdriver a few feet from us. "What is that right there, eh?"

He quickly picks it up.

"And that? What the fuck is that? Oh, wait … isn't that our X-acto knife?"

He stares at me like the question isn't rhetorical.

"Pick it up!" I yell.

We continue.

"You know how much I love our hammers, P-Unit, you little spit-fuck. Is that where our lovely hammers belong, fucko?" I point to the hammer laying on the ground by the pipe tube.

"No," he squeaks.

"You're right. You're learning. No, it's not where our lovely hammers belong. But we ain't done yet, not until the grand finale. Where's our socket set?"

I can see the gears turn in his head in real time,

then a dim lightbulb flickers on in his head. He takes off running to where he left the socket set in the mud.

I walk to him and growl, "this is pathetic. Grow the fuck up and do your fucking job."

P-Unit has no smile, no confidence. He scrambles to pick up his mess and starts putting shit where it belongs.

Every single guy out here went through this same shit. Makes you grow up fast, to understand that work isn't fun all the time, that there's a job that needs to get done. No excuses, no exemption, just get it done right.

After a good yelling, P-Unit has a few good days. He's sharp and effective, there are no tools on the ground, no garbage. But that is his greatest trick. He makes us all look right, and no one sees what happens as he moves left. This deception is his art.

The thing is that it's all my fault for being conned. The boys had all warned me, said I'd get burned, but I didn't listen.

We're working a two-man job, and I'm trusting him with tasks.

"P-Unit, I need some valves closed on my tanks. There's three of them. I'm just helping Micky here, then we'll fill the tank up with mud."

"Yup, I know which ones," he answers.

These tanks hold about thirty cubes of liquid. A cube is a thousand litres, so that means these tanks are filled with thirty-thousand litres. They have some size to them, some weight.

We start filling the tanks, and mud starts leaking all over the ground, covering us quickly to our knees. The big bosses come over, asking what's happening, blaming me. And it is my fault! It's my equipment, my blame, because I trusted P-Unit to close three valves, and he only closed two! Three-thousand litres of mud on the ground stops everyone's work and the cleanup begins.

My eyes are fire as I stare at P-Unit.

He can feel my stare and cowers like a dog who knows they've fucked up. Didn't even have the balls to admit this was his mess. I took all the blame.

Any of the other boys would have said this was their fault, would have taken the hit for their mistake. We all would have respected him that much more, because we've all been there. But P-Unit will always be a child.

After that, I started laying traps. I would leave tools laying around and flip out when I found them. I would rip into P-Unit so hard that it could be heard over the rattle of the diesel motors. Every time he did a job, even if it was a good job, I would come over to tell him it was shit and to do it again. I was determined to break him.

Later in the day, P-Unit was dropping rods. Rod after rod goes in the ground, and each rod has to be torqued to roughly three-thousand PSI, so as to not come apart underground. Because if a rod comes apart, we're fucked.

We put a thick lubricant on the threads of the rod so when we pull them from the ground, they will break apart better and prevent rust. If there's rust on the rods, it becomes impossible to break them.

The lubricant is called "dope." You don't need a lot, just a skiff to lightly coat the threads. P-Unit likes to glob it on, tons and tons of dripping, messy dope. These globs slide off the threads and make a mess, and if it gets on your clothes, it never dries, so these gobs get on the crew truck seats, doors—everywhere. It's nasty stuff, especially in the eyes of Booby Mac, who hates dope and hates P-Unit. He's bitter because the old lady hates him.

I can see him freaking out in the cab of the drill. He calls me over.

"Tell that fucking retard to stop putting so much

dope on the rods," Booby tells me.

I hop down from Booby and walk back to P-Unit. "Hey, Booby wants more dope on the rods."

So P-Unit, the innocent soul, puts the brush in the dope pail and puts more dope on the rods.

The only way to describe it is a volcano inside a hurricane.

Booby explodes, his arms start flailing in the air and, because the cab has soundproof glass, all you can make out is him mouthing the words *Fuck* and *Retard.* I can see spit flying from his mouth.

P-Unit notices and panics.

So, I tell him again, "P-Unit, put on more fucking dope. Booby wants more dope. Do what the man wants."

This kid just scoops up the dope, leaving this huge mountain of shit on the rod threads. It's falling off, making a mess, covering the drill. And our boy Booby watches this with wide eyes.

He goes frantic, losing his mind, pulling out his hair, swearing, cursing.

P-Unit is confused as to why Booby is so mad and why I'm on the ground, cry-laughing.

Again I tell him, between sobs of laughter, "more dope! Booby needs more dope! P-Unit, stop fucking around!"

Again, the poor bastard does it. There is just a mountain of dope. It's like the Guinness world record of dope on a rod.

Booby is now trying to jump out of his skin, his arms flailing around his red face. He wants to kill P-Unit.

I've done a good job. I feel pretty good about myself.

Finally, The Chief comes out. "P-Unit, he wants *less* dope."

"But he told me more do—"

"Just … put less dope on," The Chief says wearily.

Maybe it's karma, but I don't think so. I truly believe it would have happened anyway, because P-Unit is the bane of my existence.

The job is all done, and we're ready to pack up and leave. It's rig-out time!

We're told to break down the equipment and get it ready for travel. This means tying things down, putting equipment in trailers, strapping pallets down for travel. P-Unit has to help me break down my mud tanks.

Now, I don't really want to use him, but all the

other boys are busy, because all the other boys have responsibilities. So, P-Unit and I are going to bring down the walkway to my tanks.

The walkway is made from multiple, heavy pieces of steel, and we have a pulley system to transport the pieces. The railing is about a hundred pounds of steel, and I have P-Unit help me bring it up so we can lock it in place. I tell him to lock it down so the railing won't fall when we strap everything down.

As God as my witness, maybe I'm the retard. I'll be the first to admit that I'm dumb, because I trusted a little spit-fuck like P-Unit.

I'm pushing on the walkway, trying to get it strapped down, because P-Unit doesn't have the muscle to do it. I'm trying to get this steel into place, pushing and guiding the weight above my head … when I feel a crack and my vision goes black, returns for a moment, then black again.

Gravity moves one-hundred pounds of steel pretty fast. I don't have any time to react, so I'm lucky I didn't take a direct hit.

The railing smashed the side of my hard hat and sent me down to the ground, rocked to another dimension. When I come to, I'm surrounded by the boys staring down at me, and all I want is to get my bearings so I can murder a man.

I grab P-Unit and thrust him up on the tanks. "You stupid piece of shit!" I scream. "What did I fucking tell you about finishing your tasks?"

He says nothing, a pathetic look on his face.

My brain is spinning. Straining to scream is putting too much pressure behind my eyes.

I push him deeper into the tanks, screaming incoherently, louder and louder.

The boys come running, thinking I'm going to kill the kid. I honestly might have.

"Everything all right?" The Chief asks.

"This fucking cunt dropped a railing on my skull," I spit in P-Unit's face. "He's unlucky I'm alive."

The Chief separates us and sits me down.

When the adrenaline runs out, the concussion begins. I can barely hear The Chief telling the guys to never trust P-Unit, his own half-brother. I'm walking like Micky at four in the morning.

It took me three days to come back to normal, five days before my vision returned fully. Learn from me: don't trust a kid in a man's world. Find a way to destroy them before they destroy you.

The Greed

What's the point of all this shit?

I don't mean life. Though, yeah, I wonder about that, too. I mean, why the fuck am I up here? Why am I a bottom bitch to greed?

It's the greed that keeps me here. I say it's the dream, it's Ecuador, but it's the greed making me do the things I don't want to do. It's that feeling when the paycheque hits the bank account and all the bullshit, all the being gone, all the shit I sell my soul for, vanishes. Suddenly, it's all worth it. Just the simple sight of the number on a computer screen going up creates this feeling and feels like gold. Then I start thinking of all the things I can buy, all the things I never realized I've always wanted, always needed.

Yeah, it's that feeling that I make just as much as a lawyer or a doctor. I'm not saving lives or righting wrongs. I might work three times the hours, but I feel like equals, maybe even better, because my ego is so inflated.

When you have the greed, you're never satisfied. Nothing will ever quench your thirst, fill the never-ending hole. Whenever I purchase a new pair of sunglasses or a new truck while wearing well-fitted new clothes, I get this yummy feeling, like an adrenaline rush, like a big, warm glass of validation. I feel better, because I look better, because I *am* better. So, I think, tell myself—we all tell ourselves—every goddamn time. But it only lasts for a little while. Then it's old news. A memory.

The greed comes back, because the past is always better. The wanting. The never-ending hole is empty. Then back to work, slave away, my pretty sheep work harder to buy shit you don't need to fill that empty bottomless hole.

It's sad that people choose the greed over loved ones. I see it with Micky, with The Chief. I see it with myself. I'll be the first to admit I choose the greed over family. We choose work over kids, wives, friends, loved ones. We say we want to give them a better life, and we do. But, in doing so, we don't get to see them. Kids grow up and don't know who their father is. Wives find other, more reliable dick to sit on.

The greed keeps us working, and it keeps us

spending. So, we keep on working. And we keep on consuming. We keep telling ourselves: "This will be the last season, my last year. I have everything I want."

But then the greed finds another way to say, "what about this?" or "what about that?" and it's a call and a quick piss in the cup before we're back to the bush. We work and work and work, and then we get old and all we know is how to work and spend.

Would you believe me if I told you I have never seen a man retire? Not from here. It's impossible to retire because how do you go from spending with impunity to budgeting? What the fuck is a budget? *I need more!*

After that, the body is broken, the mind is bitter, and all the things we bought are our only company. No one will remember me, and no one will care. The circle of life goes on.

I remember what you have. Yes, *you*, the normal person. I am jealous of what you have. Time with your friends and family, with your kids. Weekends, barbecues, reliable sex, sports, normality.

We have the toys, the trucks, the big houses, the cocaine; we buy what we want when we want. We have the beautiful gold digger and some good women, too. We have quads, side-by-sides, and when we go to Vegas, we can and will drop two-thousand dollars in three hours, and we won't even remember. You can only dream of that.

When you're always away, you get certain freedoms. No one knows where you are. That can be nice for a certain type of person. You can get away with anything: cheating, doing drugs on the job, binge-drinking, hookers, falling apart, staring at a wall for hours instead of sleeping.

See, the funny thing is that I want what you have… I want normality. But I tell myself that you, the reliable, responsible, and oh so good person, have a dark side. I know there's a side of *you* who wants it all—the perverse and wicked and real and full. Women on top of women while they do blow off your cock. I know you wish you could tell your wife you're going to put hard dick in her ass, and you want her to be a freak … But you don't.

You goodie-goods want to taste what we have … *The madness in its purest form*.

You see, the greed is on both sides. With the greed, you can never be complete, never be whole, and never be truly happy. So, it might be best to put the bullet in the chamber, put the gun to your fucking head, and pull the fucking trigger.

Wainwright and Special Ed

We are on the Big Rig, the 220. In the world of directional drilling, it isn't big at all, but in our company, it's the biggest we have, and it's brand new. It's like a team Christmas gift.

We get to the site and set up the rig, mud system, pressure washer, rod trailer, and prep the site that we'll be living on more hours of the day than not.

Booby isn't doing so well. The battle with his wife rages, and they can't seem to find common ground. He asked her if she wants to try counselling —Booby Mac and counselling! He must really love her—and she said, "hell no!" He asked if he can do anything, and she said she doesn't know what she wants.

What the fuck is a guy supposed to do with that? There's as much logic in that as trying to tell the weather by how a dog barks. And Booby is losing his mind.

It's getting dangerous. He can't sleep, can't eat, can't focus on anything longer than a few lines of conversation. His mind is a puddle. To be honest, we should throw him in the loony bin for a month.

I have felt this pain. So have you, and you, and you. We know it's a terrible feeling, having your mind fully consumed and unable to do anything to fix it. A man needs to fix things, tinker, deconstruct, and build back anew. When they can't, failure consumes them, picks them clean.

We have a few new idiots here. Ricky is a Newfie who talks more than a guy should, almost like a used car salesman out of New Jersey. Ricky, the big talker. People who talk too much make me suspicious. Big talkers are usually equally as large bullshit artists. So far, it seems like he knows what he's doing, so I'm all right with him. If you can make it through the day without pissing me off or dropping a beam on my skull, then I can listen to your grand stories over a couple of beers at night. It's a fair trade. We also have Ed, who I both distrust and am fascinated by.

Special Ed is a forty-three year old virgin. He is one confused motherfucker. He's short and festively plump, like he had Thanksgiving every day for the last twenty years or so. He has a distinctive look on his face, like a monkey learning how to use a stick as a tool. He holds his lips open, a round *O* of confusion like he's saying, *oh how do I do this* or *oh how do I do that*.

Another thing that puckers my ass is how he can only do the simplest tasks, one step maximum. Fuelling up equipment? Easy. Squeeze the handle, fuel comes out. Ed has that on lock. Getting coffee? Ed is a master of getting coffee. It's as though evolution missed Ed, skipped a step, and left him behind. As soon as it becomes anything past a one step process, he can't make the leap.

Special Ed is guiding the hoe as we move a few pumps around. I tell him to guide the hoe so I can sling the pumps and move them to a better position. He guides the hoe bucket right into my head, just a little love tap on the base of my skull. A quick flash of the railing connecting with my head shoots through me.

I shake the memory away and look up at him. He has his hand open, all five fingers out.

"Ed … what the fuck are you doing?" I say quite crisply.

"I told him to stop, but he kept going."

"Ed, a fist means stop! Right now, you're trying to give a guy a high-five," I tell him. "Do you not know how to give direction to a hoe?"

The pipeline attracts some very special people. I am one of those people—little wild, little crazy, destined to always be a bit outside the box of the rest. But guiding a hoe bucket into someone's head is bad practice, regardless of whether or not you fit in at dinner parties. If you don't know how to give signals, then ask, and we will teach you. Special Ed isn't much of an asker.

"Ed, are you fucking retarded? Or just a lazy sack of dog shit? In all my years, I have never met anyone like you."

"Hey, man, you're the same as me—you're just a worker like me. You can't tell me what to do!"

He thinks we're equals, that I'm just a labourer like him, that I'm a Neanderthal like him.

I see red, more than angry to be compared to this filth. I give him his first test out here, animal to animal, to see who is dominant, to find a weakness that I can later exploit. The bush is the western world's last coliseum.

Don't feel sorry for Special Ed. This happened to all of us.

"You worthless, motherfucking, lazy, fat fuck!" I yell at him, my arms in the air.

My tribe knows my battle cry. They know Eddy's

head is on the sacrifice stone. With thunderous words as the bludgeoning hammer, about to smash him towards the pipeline God's. People gather a little closer for worship.

"You listen here. There's a chain of command, Ed. Now, the driller over there, he's number one, then it's me. You're at the bottom, bitch. You're less than the whores we fuck. If you have a problem with that, either do your job or file a complaint in the I-don't-give-a-fuck box!"

There's a small silence after my explosion, so I continue, "and put your fucking tools away, you puke."

I can see his eyes watering up. Pathetic.

I gave him a chance.

I know what you're thinking … I'm a bully, a bad guy, a cunt, a real motherfucker, and if I ever did that to you, you'd knock my head clean off. And you're right; I'm all those things. But don't be so sure you'd fight. There are only a few people on earth actually willing to stand tall and fight back. Shit, I respect those people. This is life out here. There are rules and laws, unwritten but set in stone. Don't take it personally. It happened to me, and it makes you tough.

Poor, poor Special Ed. He told all the boys on the pipeline that I yelled at him. When he told them, he was almost crying, which only made me look like a super star.

"Why did he yell at you?" Booby asks.

"Oh, because I left tools laying around. You know he's just a labourer like me; he's no better."

"What?" Booby snaps. "You do whatever the hell he tells you to do. You're at the bottom."

"What? No, he—"

"And put your fucking tools away! I would have done the same."

The boys hate a rat. Hell, everyone hates a rat. If you ever decide to come out to the bush, keep it to yourself. If the boys are hard on you at the beginning, it's to test you, to see if you can handle criticism, joking around, the roughneck shit we need to do to get through the grind. It's to feel out if you can handle being one of us. If you decide to rat or tell people that guys are yelling at you for not doing your job, well, you just painted a target on your back that everyone is looking to hit. Every guy across the site is lining up to be the one who drives the final nail in your coffin. We get bored out here, need entertainment. No one feels bad when you rip on a rat.

The crazy thing is I don't feel like a bad person. I feel like part of a crusade to seek out the shit that people bring with them, a lethargy and demand to be pampered and made special. I feel part of something larger, filled with a passion for only the best, the craziest, the maddest of people surrounding me.

Booby Mac

My dear Booby Mac, truly, my heart goes out to you. You're a great guy. Always lending a hand. Genius about drilling. To be honest, the most knowledgeable guy I know. You're a class-A pipeliner, ready to work long, hard hours to complete any job.

I've seen your bastard ass work literally three days straight. I've worked with you for three seasons now, and you could almost call us brothers. We've been through hell and back. I've known you at your best … Oh, but I've known you at your worst. I've seen your strong will brought down to its foundation, to its lowest point, a crumbled shell of a man, all insanity and madness and bad ideas.

At the beginning, you and I didn't really drink. We were strait-laced and clean shaven. We worked our hours and made good money, living the life everyone imagines they'll live before they get up here. Then came the recession.

The world loves to drive cars and trucks, eat

exotic foods, sport hunt and fish for dolphins and whales—which are mammals, by the way—and they love to be warm during the winter. But, for some strange reason, the world hates oil and gas. It's confusing.

See, the world loves planes, uses them to travel so they can see beautiful architecture they weren't blessed to be born near, but the world hates oil and gas, hates pipelines, hates pipeliners. So when the bubble burst, everything stopped. All the pipelines, production, building, trucking, manufacturing, and everything else. Basically, our whole fucking world.

With that recession, good old Booby Mac, your company, Lilliedale Drilling, couldn't make the payments and was one of the many companies that went under. Bankruptcy. Your dad, sixty—retirement was right fucking there—he took the bankruptcy. Lost everything—his house, his trucks, his hats, socks and shoes, the soap he washed himself with. But you stayed strong, Booby Mac, you stayed strong. You didn't get angry or mad, not outward. Maybe a few extra drinks after work, but who could blame you? I couldn't. Middle finger to those who could.

Then your Mom got sick with cancer. The bad cancer. The cancer that kills. The cancer that poisons, drains away the woman who loved you since you were a cell. The cancer that takes away the woman who made lunches, dinners, and put Band-Aids on skinned knees. The kind of cancer that took your Mom away, that's what kind she got. To a better place, we say, but reality is that she's in the place where worms feast.

You had a cry, because you knew that in life there was death. She was in a lot of pain, and it was better for her to surrender to the cycle of life.

Oh, we are not done yet … Oh Lord, we ain't done yet.

After your mother died, your little brother, all six-foot-eight inches of him, came out of the closet. Your only brother is a rope sucker, a real French fairy. Sorry for the code, but what I'm trying to say is that your brother prefers the company of men over the company of women. Sexually, I mean. And your dad didn't like that. Not one bit.

His heart started malfunctioning, ticking when it should tock, tocking when it should tick, and beating so fast he would turn white when he moved his head too quickly. But he survived! You felt good about that. The darkness was over, like the tunnel was ending and light was coming.

Then the final straw. The one thing you couldn't take, Booby Mac, was when your wife started playing games, when she started to resent you. When she stopped returning phone calls or messages.

When a man tries to understand what's going on in a woman's mind—"why does she suddenly hate me?"—he starts to lose his mind. Tries to use logic to figure out one of the world's greatest mysteries. The relationship dynamic was the same for seven years:

1. Work out of town.
2. Come home.
3. Party and fuck like rabbits.
4. Back to work with some of the summer off and holidays.

Nothing had changed! Why was she being such a bitch?

Men try to look at it with logic, with reason. "Okay, if I go home, I can fix it." Kind of like a broken sink, right? We say things like, "I won't work out of town, I'll call more, I'll do what you want." But the damage couldn't be repaired.

I saw what madness looked like. I saw it in your eyes. You kept running the scenarios in your head: you cast aside to the wolves in the cold while she was warm and comfortable with a hard dick thrust inside her, reading off a list of your greatest embarrassments.

So, Booby Mac, you have not been doing good. You've barely been sleeping, scarcely eating more than nibbles. The stress is melting the bulk from your frame. You're living life on a roller coaster. One day saying you two are figuring it out, you see her point of view, and then, not twenty-four hours later, you're stomping around camp, wrecking people's nights, saying, "that fucking bitch, what the fuck! Bunch of people at my house, drinking my liquor."

You're drinking more. A lot more. I think you've come to the realization that there might be another man, but you can't really come to terms with the idea. You fight with the idea and put it away in the dark of your mind.

Booby Mac, my heart is filled with your hurt. I wish I could help you. Give you a magic pill to make you feel sane. Fuck, I'd even give you a hug. My dear friend is in pain. My dear friend suffers. Only the God's laugh while the rest of us are here, part of the show.

You've started having dreams that feature me fucking your wife. I have to say, I do not like this development. This is not good. You told me about it.

In the dream, you walk into your room (except it's a dream, so it's not your room, but it is? You get it), and there's me, just fucking your wife. The lights are on. All the lights. Bathroom lights, television's on, lamps, everything. It's like you have snow blindness it's so bright. There I am, your wife bent over and just going at it, on your bed. My white, veiny cock going in and out, balls slapping against her cheeks. I mean, really giving it to her! Then I look at you and say, "Don't worry, Booby; I won't be too much longer. Then you can have her back."

When you wake up from the dream, there I am, sleeping peacefully, like an angel, in my hotel bed. Sleeping like a beautiful baby child blessed by the Pope himself. Not making a sound. Yet, when you see me, you want to kill me. Strangle the life out of me because I violated your wife in your dreams. Doesn't that seem a bit strange, Booby Mac? *You* made it happen!

After, you can't sleep well. Stewing, brewing with hatred. You hate me, and you hate your wife, but I think we both know who you really hate.

The alarm goes off. Time to work. I wake up well rested, but you haven't slept. You've turned into what I would call a wild psychopath's lunatic, filled to the brim with bile. A man filled with so much rotten that his eyes are red. I'm half-asleep on our drive to work, my head bouncing stupidly off the window before nodding back against it. I didn't know what you dreamed about the night before.

"I'm going to fucking kill that bitch, bro. Fuck that stupid whore!" you scream like a banshee.

My eyes snap open.

"Fucking bitch. I'm sure she's fucking another guy. Hanging out with different people, inviting strangers into my house. To *my* house! For fuck's sake, *you* were fucking her last night!"

I look over at you, slightly more confused than scared.

You stare daggers through the windshield. I can hear the plastic of the steering wheel squeal in your grip. Your face is wild with bad ideas and poisoned thoughts.

"Uh, what? I was beside you last night."

"In my dreams! You were fucking her. You and my wife, last night. I haven't slept all goddamn night."

I'm a little nervous about the news that I fucked your wife. I don't trust the irrational, but they usually come in two types.

There's a saying with bears. If the bear is brown, get down. If the bear is black, fight back. Now, Booby Mac, what kind of bear are you?

"You gotta relax, man. Dreams are nothing. You're mad about something, I see that, so let's just chill and figure it out."

"It's too late for figuring things out," you say. You've got insanity in your eyes. Why didn't I drive? "If she wants a divorce, I'll give her a fucking divorce. I'll fight her for every last cent."

You're still screaming. Your foot is heavier on the gas. "I'm going to call her right now. Tell her I know all about what she's up to in my goddamn house."

There are a few things running through my mind. First, why am I in your dreams, fucking your wife? It's dangerous fucking another man's wife, even in dreams. Second, I can't let my dear friend call his wife in the heat of anger. It shows weakness.

"Booby …" I begin. I pause and chew my tongue, let the silence marinate. "How many times has a woman woke you up angry about something you did in a dream? Don't you find that a bit ridiculous? I mean, it's a dream. It's not real."

You bite your inner cheek, look at me out the

side of your eye, then back to the road. "It's not the same," you say.

You're a black bear, Booby. And I'm going on the attack.

I strategically plan my attack. I'll wait until we're close to the work site, and then I will strike. Steal your phone, and because I'm much faster and stronger, you won't be able to get it back and you won't be able to call your wife. This will give you time to calm the fuck down.

As we pull into the site and the truck starts to slow down, I quickly jack your phone and jump from the moving truck.

I can see you contemplate why I jumped. See you thinking it was a bit funny. For a moment, you're calm, civil. You're Booby Mac, the one I know. Then I see your eye twitch and your mouth quiver and realization flashes across your brown eyes.

"Give me the phone," you say, anger seeping through calm.

"Booby … you got to calm down before you call Cathy."

"Give me the fucking phone."

"I will …" I say, "in an hour."

You attack me in a sprint, long legs pumping like

you're running from a monster. You can cover a lot of ground in a short space, but I am graceful like a mountain goat, determined and nimble. I dive, dip, and dodge to avoid you.

The pursuit begins! It's a classic game of chase. Booby, you're the same age as me, considerably taller, but far less athletic. It feels good to be a kid again, playing a furious game of tag. It used to be jungle gyms and swing sets. The biggest sin was broken dishes. Now we're running around million dollar pieces of equipment, playing tag over a phone.

You're determined. We're circling the drill like it's an Albertan merry-go-round, you're cursing my name, calling me *cunt, bastard, quick little bastard.* You pause to catch your breath, then some more *fucks*, *cunts* and, finally, *bastard* again. We're having a laugh about it, but soon your anger boils over and the game comes to a halt. Your persistence won my heart, and I reluctantly gave your phone back.

As I watched from my perch on the reclaimer, I see my dear friend throw his hand in the air like Zeus throwing thunderbolts of rage. I can hear your screams over the howl of the diesel motor. Your face red with a dying marriage, your heart boiling with hurt and sadness. The phone call was short and pointless.

I know you're hurt. If I could, I'd give you a hug. Or gift you a time machine to fix the past or move to the future, far away from this feeling. I see you go to the outhouse to cry silently, uselessly. Try to vomit adrenaline. Vomit all the anger and frustration. You should let yourself cry, Booby Mac. But you can't. Definitely not out here where it's cold and bitter and there's no room for kindness or weakness. Only the strong survive. If there's weakness, build a wall to conceal it. Store it in a jar on a shelf inside your chest. Out here, it's the pipeline and only the hard and the mad can make it out of here alive.

After the Wainwright job, you had to take some time off. You didn't say why, but why waste words? Your wife was leaving you and your life was leaving with her.

Ol' Booby Mac, you're leaving us to go on the most courageous, depressing bender a man who was to be king ever did to himself. God bless you and Godspeed, Booby Mac.

Too Much Time to Think

With Booby gone and the Wainwright job done, there was no time to rest and I was shipped out to another crew to help out for a couple days.

These guys, I would call normal. Normal-ish. Pipeline normal in that they're just weird, not borderline destruction in person form. I know a couple of the guys, like Rippin' Ralph and the Cole's.

Rippin' Ralph is a cliché. He is the five-foot-nothing rig worker, driving a jacked-up Chevy truck. When you see him hop, literally hop, down from his driver's seat, you make a penis joke, and he hates it.

The Cole's are actually one person, named Cole —but wouldn't it be great if they weren't?—who talks to himself a lot. Full-blown conversations with himself in two voices, one low and one high. It's weird how quickly I got used to it. I'd hear him talking away to himself in the drill, all day long.

"Well, if the liberals get in we might be out of a job. The liberals hate oil," Cole One would say in his low voice.

"Then we could go back to school. We've wanted to do that for a while," Cole Two would reply in his high voice.

"Let's not be hasty," said Cole One. "What we can do is work as long as we can, save some money before going back to school before they shut down oil completely."

"Could they ever completely shut down oil?"

"Of course they could."

This would go on for hours. Like I said, normal-ish.

My first night out with the Cole's, he told me a story about having sex with a girl who was a deaf mute. She couldn't hear or talk, but the Cole's spoke enough for the both of them. She made lots of sounds, grunting, yelping. A pretty strange introduction, but I guess deaf mutes need a good fucking, too. And honestly, I like the Cole's. He's just so weird it's hard not to smile.

Aside from the two characters, the boys are pretty tame. Wake up, the boys gather in a semi-circle and blaze one down, work all day, then back home to sleep. It leaves too much time to think. No more of Micky's constant consumption of booze and temptation. No Booby going off about his doomed relationship, sitting in the truck, alternating between figuring out how to make her love him and planning her murder. Nothing was even breaking down to keep the mind busy. No idiots, no kids, no drama. No fun, really.

My mind has been weakened from working forty days straight. I get less of a choice on what thoughts creep in.

I close my eyes and find myself in the past, Nami coming back with force, all the repressed thoughts breaking through and flooring my mind. I thought I stuffed that shit down, suffocated those emotions enough to kill and bury them.

My darling Nami. Such a sweet, lovely, tiny Japanese lady. She was a lady in the streets and a freak between the sheets. And that's exactly what I want. A lady to show off to family and friends; a respectable, delicate, intelligent lady. In the bedroom, though, it's another story. In those moments, we forgot all the fights and tears and pointless bickering. The good times are heightened in memory. She fit so well in my arms. It felt good to protect something, and she was mine to protect. We would be together for hours and not say a word, not feeling the need to do so. The beer and food were more delicious when I was on dates with her, colours were brighter.

The universe is a weird thing. I can't tell you how it works or begin to explain its method. I hadn't taken time to think of Nami in so long, and during all that time, not one word from her.

Two days after the dam finally breaks, she messages me. It's as if I wrote it. It's filled with emotion. I can feel her tears, her heart as broken as mine. I read it again and again and again and again …

New Crew

Wilton was making love to a cigarette while we waited outside the shop for the news of what was going to happen with us. There were two broken crews, and we were the broken pieces. People come and go. We're all sluts of the pipeline, and whoever pays us more is going to get the workers, and a few sluts found a new pimp, so Rossko was figuring out where to fit us broken pieces.

"Hey, boys," Rossko says, stepping out of the trailer. "What's the difference between a pickpocket and a peeping Tom?"

"What's the difference, Rossko?" Wilton says, blowing smoke.

"One snatches your watch"—his laughter fills the air before the joke is complete—"and the other watches your snatch!"

Wilton and I chuckle politely.

"Well, boys, I know where I'm going to put you," Rossko says.

"What's our fate?" I ask.

"You boys are being paired with a young guy by the name of Cole," Rossko says.

"The guy who talks to himself?" I say. I miss that guy.

"Different Cole," he says.

Damn.

But Wilton and I give each other a sinful grin. Working with my best bud would be lovely, even if we didn't know who this Cole was. Time would tell if he would make it or if we would chase him away.

"Boys, do you know how to stop a dog from humping your leg?"

"How do you stop a dog," Wilton says, stamping out his cigarette, "from humping your leg, Rossko?"

"Pick him up and suck his dick!" He howls with laughter. "Anyway, get your shit together, because you're leaving in two days to Grande Prairie, working for Orange County Pipeline."

We had heard about the Orange County boys. They were cowboys, wild ones. Low safety regulations and a lax attitude towards the environment. Works for us.

The only question was: who is this Cole? Is he a lazy punk? Or is he going to be one of the boys? Our questions are answered two days later when he strolls up to the truck, head cocked to the side, while Wilton and I are waiting inside. He's twenty-one years of insecurity, atop long, lanky legs. His arms hang down nearly to his knees.

Wilton and I give him a once-over and shrug. I guess he's the final piece to our broken crew. Now we are complete.

BSCB- An Adventure for Chiggins

Part One

The Orange County boys are Grade-A gold with full-on don't-give-a-fuck attitudes. Just as we heard before, these boys seem on a vendetta against mother earth. We couldn't have fucked the environment quicker if we shoved our dicks in an ant hill. One day in, and we break down, the blood of our rig smeared over the purity of the earth, seeping into the soil. Young Cole is on clean up duty, scraping away at black globs of dirt the consistency of industrial glue and depositing them into bags and pails.

Wilton calls our mechanic, Rory. He holds up fingers to me while he's on the phone. Three fingers … four fingers … back to three … Three fingers, three hours, for sure. He hangs up and shrugs.

"Looks like we're making a beer run," I say.

Wilton turns on his heels and starts for the truck, while me and Cole kick dirt over the spill.

Wilton always has a sweet tooth for beer runs

and gas station chips. I think he likes the solitude of the drive, however momentary.

The Orange County Boys play a vanishing act at four-thirty and are long gone, leaving Cole and I in the wilderness of Spruce tree silence and long, blue skies.

Cole doesn't speak much more than a mumble. He keeps his head down and works, following directions. He's really skinny, so much so that I wonder how his legs support his tall frame. He has wide, adventurous eyes, always watching, assessing. I think he understands more than he lets on, a wallflower type with a whole internal world churning inside his head. It's just the two of us, and the only thing I get out of the guy is that he's from Maidstone, Saskatchewan and that this is the farthest away he's been from home in five years.

Staring over his shoulder into the distance, I find myself needing to ask an important question.

"Is your mother your sister?"

"No!" He laughs awkwardly.

"You sleeping with your sister? 'Cause I don't know if we can trust a sister fucker, weirdo."

"I don't fuck my sister," he says. "But I do fuck other people's sisters."

I like that. That's a good answer.

"So, you are a sister fucker!"

"Oh shit, I guess I am," he says. "Why? Do you have one?"

"I do, I do. But he's got a dick. His name's Marko. Have at 'em. Even Marko needs love, too."

Wilton arrives in a cloud of dust and emerges gloriously with the beer, and Rory.

Rory is pure class. Quick to a joke and to light up your smoke. Rory is too smart to be working out here with us, but he stays season after season.

The bottles are a delicious sight. Not only did Wilton bring a case, but so did Rory, and the bottles have beads of sweat running down the glass, begging us to open them. We humbly obey our masters.

Watching Rory work is art. He can paint the corners like Rembrandt. He flows flawlessly from one bolt to the next. His brain is a problem-solving machine. He knows the exact size of each bolt, how many there are, and every trick to make it happen fast. He barks orders at Cole to get him tools during his breaks to sip beer.

"Hey, C. Higgins, put this back in the right place," he yells. "13mm goes in the 13mm spot, like those puzzles you did last year in kindergarten."

Cole runs and follows orders like an obedient dog.

"C. Higgins, can you—wait, C. Higgins… Chiggins! That's what I'll call you, Chiggins," Rory says.

We look at each other for confirmation. It fits flawlessly.

Just like that, young Chiggins is born. Another pipeline birth.

With a guy like Rory, you have to earn his respect; there's no other way. When I first met his cunning, cocky self, I would call him a cunt, straight wizard sleeve. Brazen and coarse, but if you can work, he quickly warms up.

Cole is in the brazen and coarse stage. The worthless stage.

"Hurry up and grab me that pin bar."

Chiggins runs and grabs it like a good dog.

Rory quickly gets the rig fixed and ready for battle, downs his beer, and tells us he knows a good spot to eat. We follow Rory's lead to the strippers.

The smell of stale beer and strong, flowery perfume stings and confuses my nostrils. There's a guy at the bar who has been sitting on the same stool for twenty years, ordering the same beer. He's not a customer; he's part of the place. He's baked into the walls. A relic.

Chiggins sits down in Perv Row and Wilton, Rory, and I take a table.

"How's Chiggins?" Rory asks.

"Too soon to tell," I say.

The show starts and beautiful long legs of low self-esteem twirl on the pole.

Wilton's eyes are fixed on smooth, bare ass as she works the stage. He's in love.

"I raise a toast," Rory says, raising his bottle. "To all the bad fathers and broken homes. May they never rest, never stop doing God's work."

We cheers.

"We're still on the clock," Wilton says, smiling.

"We don't get paid enough for this," laughs Rory.

Chiggins, lanky, wide-eyed bastard, is the only guy sitting in Perv Row, and we watch from our table as he chats up the bottle lady. It seems like he has her captivated, but the only people who like a guy more than waitresses are strippers, so we can't be sure.

The stripper is up on stage with all her glory hanging out, trying for his attention, but he's barely giving her a look. I understand we're at a strip club and that this is a job, a job to make idiots like us give our money for a chance of blue balls, but Chiggins is moving different in here.

He saunters over to us and sits with a smile. "That bartender gave me her number," he says. "She invited me to a party this weekend. You guys wanna come?"

"What?" I laugh. "She just walked up to you, talked to you, and gave you her number?"

"Yeah."

"What did you say to her?"

"Nothing. She just came up and started talking to me," he says, shrugging. I don't think this is the first time this has happened to him. "I mean, I said I'll come party, for sure. She was practically begging."

The bottle lady comes up behind Chiggins and hugs him from behind. He looks awkward, sitting there and waiting for it to end. She has crazy eyes. I like her immediately.

Young Chiggins surprised me. A man in mud-covered clothes who talks so quietly that he can hardly be heard pulls a number at a strip club just by existing… He might be worth keeping around.

Part Two

When I'm right, I'm right. The bottle girl was bat-shit crazy. We're talking stage five nuts, fake an emergency and run, take your condoms from the trash kind of crazy. We've taken to calling her BSCB (Bat-Shit Crazy Bitch).

BSCB messaged Chiggins forty-seven times today, and his phone vibrated like a bat out of hell atop the truck dash while he looked to us for advice. And we tell him to keep it going, don't let it stop. This is like oxygen for us pipeliners.

Before we arrive at the bar, the boys find a back alley to buy a few white bags of enlightenment. Then we stroll on by the slutty bunnies, high-five the Trump look-alikes, walk past the dirtbags who didn't care enough to dress up … wait—that was a mirror. Shit, we're the dirtbags.

BSCB, flanked by her stripper friends, like a general's honour guard, spots Chiggins and attaches herself to him. She lures him away to overwhelm and suffocate him, and we're fine with that. Good friends are the ones who sometimes let you suffer.

There's a gentleman sitting at the head of the reserved table where BSCB and her friends lead

Chiggins.

He looks like a Picasso painting, his face darkened with tribal tattoos. He's fat, bald, and looks mean as hell, yet he has a double bubble chick on each leg with his arms around them. He commands the table.

When Wilton and I say hello to all the strippers, he doesn't even look at us, like we aren't worth his time. We're just two flies buzzing around his shit.

"Wanna go play the VLTs?" Wilton says, hoping to get away from the man. Wilton loves everything that's bad for you, a man for the pleasure of sin. He's the devil on my shoulder, and he's the best a guy could ask for.

"After shots," I say. "Little liquid luck."

Whiskey straight, twenties in the machine. We bet together so we win together. When we lose together, we grab more beer together.

Gambling is a brotherhood. Wilton's a max bet kind of guy in all aspects of life. He needs the best house, the truck with all the options, the nice clothes, newest phone. If he's losing his hair, he'll find a way to get some more at any cost. First bill, max bet, mermaids match, and we win big.

Wilton is shouting, hooting and hollering with

the *ding-ding-ding* of the VLT. I'm riding an invisible horse around the bar, celebrating one-hundred-and-fifty dollars right off the hop. We cash out a bill, grab Chiggins and BSCB, and fill everyone up with another round of beers and shots.

I start to listen to BSCB and realize she could be the most upfront, crazy girl I've ever met, which I think is pretty cool! Most people, they try to hide behind a mask and lie to the world about something so fundamental as who they are. If you're nuts, you're nuts. If you're boring, then be boring! Be you! This girl was naked, no mask, completely and honestly nuts. It was refreshing.

Our second trip to the VLT goes more as we anticipated, and the machine eats our cash before we knock back a quarter of our beer.

Wilton gives me a wink and heads to the bathroom for a prairie pick-me-up, so I head outside for a quick dart.

Wilton starts charming, schmoozing the smoke pit crowd, brings two girls with plain, pretty faces over to meet me. He talks and compliments, jokes and charms, and they're hanging off him like a baggy sweater. He has a girlfriend, but hey, talking isn't cheating.

End of the night starts rolling around, and we finally see Chiggins emerge from the clutches of BSCB. She was on him all night like his shadow. He looks like a photo of young men returning from 'Nam, staring dead-eyed into the middle distance.

"Even when I went to the bathroom, she waited outside," he says to us, his eyes begging us to help.

"True love is a wonderful thing," Wilton says.

"Honestly, I'm jealous," I add.

Chiggins shuffles his feet, looking over his shoulder at BSCB, who watches like a hawk. "I guess there's an afterparty," he Mumbles, "but only I can go."

"We can't come?"

"I already asked. They said *no*."

"Where?" Wilton asks.

"It's at the house of the guy at the big table. With the face," he says.

"Bro, you have to go to this party and tell us how it is," I say. "That guy has definitely killed a man."

He looks at me like a puppy dog.

"But you're his friend! You don't kill friends," I say.

"I know," Chiggins says. "I'm still scared."

"Ah, you only live once," Wilton tells him. "If it

gets too weird, just come back to the hotel."

He doesn't seem convinced, but he doesn't have much of a choice. BSCB snakes up from behind, wraps her arms around his skinny waist, and starts dragging him backwards into a limo, waving at us as they go.

Chiggins waves half-heartedly, and we bid them farewell with big smiles plastered on our faces, giving Chiggins the thumbs-up.

We watch the limo pull away from the club and down the street. Wilton lights a smoke. "So, we're gonna need a new guy for the crew, eh?"

"Yeah, he's definitely dying in that house," I concur.

Part Three

It's clear when the entrance to the house is blocked by an iron fenced gate that needs to be opened by speaker box, Chiggins is out of his element. He's a raw form of intimidated and drunk, and the surprise on his face and innocence in his eyes makes him a magnet to ladies.

Most houses don't have pillars by the front door. Most houses don't have doors that belong to a giant in a fairy tale. Most don't have the host offer you a pill upon entering while saying, "here Alice, take this pill. It will make you bigger."

The interior even smells expensive, a combination of oak hardwood and Cuban cigars.

BSCB takes him to the living room where women are dancing slowly with their titties hanging free, powered by the smorgasbord on the kitchen island. There is everything under the sun: uppers, downers, blow, sidewinder, special K, unmarked pills and, of course, liquor: a lot of it. Chiggins and the owner of this house are not the same brand of person.

BSCB opens the fridge door, pulls out a Mio squirt bottle. "This is GHB," she says, shaking the bottle playfully. "The date rape drug." She squirts some down her throat and winces. "Oops … Now you'll have to rape me."

The owner enters the kitchen and sees Chiggins in work wear, confused and nearly scared. His tattoos are somehow more confrontational in the harsh, fluorescent light of the kitchen. "Who the fuck are you?"

Chiggins shrinks in the presence of the man.

"He's my friend, Tony," BSCB says in a slurred voice.

Tony examines Chiggins, susses up his past, present, and future. He sees his issues with his father, his inability to say no, and sees how to control him. "You ever try GHB?" he says. "Fucks the girls right up, but they love it."

Chiggins shakes his head.

"Drink it."

"Okay," Chiggins says. His hand is shaking as he takes the bottle and squirts some into his mouth. It tastes like a mouthful of ocean water.

Tony grins, revealing a mouthful of crooked teeth. "And that's all there is to it."

For the first time all night, Chiggins is free from BSCB's grasp. She's in the living room, the only woman wearing clothes, doing the hoedown throwdown shuffle.

Chiggins eyes the mountain of cocaine on the kitchen table and quietly, respectfully, divides a line from the pile and inhales deeply. His nostrils sting, and his teeth go numb. He grabs a cold Heineken from the fridge and starts to wander.

There are gangsters talking shop surrounded by naked women dancing on a medley of different drugs.

Chiggins is moving with a solid stagger, balancing the GHB with the coke somewhat nicely.

He walks up a spiraling staircase, two-handing his Heineken and forgetting he needs to drink from it, while he admires paintings along the walls.

One catches his eye. It's a woman broken into pieces. Her hand in the top left corner, her tits floating like clouds, legs half-hidden behind a bush held up with sticks. He looks back at the girls dancing, his eyes constantly refocusing, recollecting where they are. Chiggins keeps moving upwards through the house, hoping for something to make sense.

He opens a door at the top of the stairs. It seems like it's a bedroom, but it resembles a dungeon. There's black leather along the interior, chains against the wall, and slings hanging from the ceiling above the bed. On the bed is a woman riding a man like an animal, begging for him to give her more cock, fuck her harder. Chiggins could see a vein in his temple pulse as he did what she asked. He shut the door out of respect.

He went back to the kitchen, content to sneak a few free rips and be on his way, but curiosity seemed to get the best of the naked women.

"Who are you?" one asks.

Chiggins Mumbles, "I'm here with"—he searches around the room and sees BSCB facedown on the couch—"her." He points towards her lifeless body.

"And she just left you all alone?" the woman says, pouting and stroking his face. "We'll take care of you."

They bring him to the hot tub, strip him down, and join him in the water. He feels like a king.

He leans his head back, brain dancing with the events of the night, slipping into the drinks and drugs, before waking the next morning beside a woman he's never seen before, suddenly aware he needs to get the fuck out of whatever wonderland he found himself.

Cocaine and Tim Hortons

If you ask a politician, they'll say the driving force of the pipeline is innovation, perseverance, hard-work, resilience, and maybe a bit of the ol' Alberta advantage. If you ask a pipeliner the same question, they'll tell you the truth: the driving force of the pipeline is cocaine and Tim Hortons.

Out in the bush, you never know what kind of shitstorm will be thrown at you, what type of break down will happen, or when you'll need a part or special tool to get the job done. Of course, there is never enough time, so it's always a panic to get whatever item we need for the job.

Truth be told, working a twenty-four hour shift is routine. The most I worked in one shot was sixty-three hours, and that was one fucked-up trip near the end. I would be at the wheel of my truck and arrive at the next destination with no memory of leaving the last. On paper, we say this never happens, but the reality is we do some long days, and sometimes they stretch into multiple days at once. We do whatever it takes to get the job done.

We're not robots. We're no different than anyone else. Flesh and blood and skin and bones. Our bodies start to break down, our minds start to fail, but we still need to keep them awake. We need to keep them electrified, jolted, revving at a high RPM, even when our eyelids start to close against our will. Whether it's driving an all-nighter to get the proper tooling or finishing the last bit of a pipe pull that drags into the wee hours of the night, sometimes you can't plan for life. What should have been home by six, turns into an all-nighter, and back on site for seven a.m. Yes, a guy needs to kickstart his heart with a strong cup of coffee and a sneaky line and loud tunes to keep the drive alive.

But let me be honest, these men are not angels, not even saints, definitely not repenting anytime soon. Some have got the devil on both their shoulders. It could be any day of the week, even the holiness of Sunday, when something inside stirs, a yearning comes from inside, and they're out in the bars, making jokes about how they don't like cocaine, but they love the smell of it.

Cocaine on the brain will make a guy drive three hours, roundtrip, to Edmonton to meet up with his dealer and buy a moment of comfort, to be complete. Cocaine will help you lie to your wife or girlfriend and tell them you only do it "once in a blue moon." It'll wear you down. You'll turn down a line to start the night, say the vibe isn't right, but after one beer, maybe two, you're texting your dealer for a bag so you can feel atop Mount Olympus, next to almighty Zeus, a God of the Grande Prairie Holiday Inn.

Cocaine will be your best friend when you need to stay awake and drive all through the night. It'll help you talk shit all through the night until the sound of the birdies in the morning. Talk about how you'll travel the world someday, or fuck the cute bartender, or how you'll never do cocaine again, or how you feel emptier now than you did as a kid.

Another call is made for more until you are hopeless, unsatisfied with yourself, and decide to make an attempt at sleep, thinking to yourself, "tomorrow is going to be a tough day. Why did I do this? Fuck my life." Which is all the more reason that Timmies is our lifeblood.

When you're feeling like a sad sack of shit because you were railing lines until the start of your shift, Timmies doesn't judge. It looks you square in the eye with respect, gives you a shot in the arm, and tells you to face another day, you goddamn beautiful disaster. Tim's is a close friend to us all out here. If you want to make friends with everyone on site, bring a dozen donuts and a box of coffee. You'll be loved by many, even if you're still nursing that hole inside, waiting for the end of shift to buy another little bag of good intentions.

Driving

Chiggins is young and dumb, but we've decided he's one of the boys. His innocence is seductive. He's fun to corrupt. He's become a little brother for us to guide. He's a man of simple tastes: smoking, cocaine, driving drunk, and pussy. His quiet demeanour and nearly inaudible mumbling makes him even more endearing.

We're finished up with the Orange County Boys. Pipes are in the ground, diesel has been burnt, and money has been deposited in our banks. End of the job means end of the job beers, and word spreads fast through the bushmen grapevine that celebration beers are taking place at the strippers.

Wilton and Chiggins get hooked up through a worker in another crew and meet a high school kid named Pablo selling half-grams behind the Wendy's. They break it down on the dash of the truck and take turns offering the first line to the other, graciously turning it down and sending it back to the other man.

Once we enter the strippers, Wilton and Chiggins

dash to the toilet, ungraciously shoving the other out of the way to be the first to use the stall. Not all drugs are alike, and the boys have been lucky enough recently, but Pablo was selling stuff stepped on and mixed with baby laxative, so the boys will be a bit before they're ready to emerge. They'll still do more lines, too.

I lean against a pillar, waiting for more guys to join, saying *hello* to every girl who walks past me. The same regular from before is here, still sitting in the same bar stool with the same beer. A few more wrinkles, a few more stories, a few less bucks. I wish I didn't recognize him. I wish I didn't recognize me in him.

The night is routine, and we hate it. We talk bullshit, drink beer, do lines. The strippers dance, we give, they take. Beers go in, lines go up, money goes out. We feel good. No, we feel great. No, we feel like God's. No, we are higher than God's: God's on clouds. No, we wake up, we feel bad. No, worse. We are scum, we don't have any money, we don't have any drugs. We have less. No, we are less.

The drive home is fucking terrible. Three separate vehicles, playing leapfrog down the highway, emergency stopping at different gas stations to shit and vomit. Why do we do this to ourselves? Why do we poison the body with this artificial pleasure? Is constant pleasure the same thing as happiness?

Wilton and Chiggins are worse than me. They didn't sleep. They're sweating vodka and shooting blood from their nose into tissues. They have a hard slouch, and their faces wince with pain after every step. Our breath is toxic, eyes glowing red, the sweat from our pores leaking what we consumed so strongly that our vehicles smell flammable. The smallest things put us in a rage.

When we get back to the shop, we all dive into our beds. Not even the dead sleep this well. Then I arrive at the shop feeling refreshed and ready to go, waiting on the bed of my truck for Rossko to arrive.

I've been thinking about that man in the strip club, seeing the same sights, tasting the same beer each night. I can't shake him from my mind.

Twitch

Rossko calls me and Wilton over and launches into a journey. We hold on for the ride.

"So, there's this couple driving down the road. The rain is pissing down in sheets as grey as lead. Visibility is shot—nothing. All of a sudden, they slide off the road and, sure as shit, they are stuck tighter than a bull's ass in fly time.

"So, the guy gets out of the car and tries to push them out of the ditch. She's mashing the gas, he's putting his back into it, but fuck all happens.

"So, Buddy stands there in the rain, soaked through his shoes, and sees a light up the road a mile or so. He starts walking through mud up to his knees, rain's still pouring; this guy is fucking miserable. Finally, he gets to the light, and it's a farmhouse.

"So, he walks up to the house and sees the front door wide open. He reaches in and pounds on the door, waits a minute, bangs on it again, and still no one comes, yet he hears something inside. He steps inside to have a look, and there is this naked woman standing in the middle of the living room, yanking on her tits. And there's a guy sitting in the corner, pouring a pot of water on his head while he's jerking off."

"Amazing," Wilton says under his breath.

Rossko leans farther toward us, getting into the story. "So, Buddy turns around and gets the fuck out of dodge, nearly sprints back out into the rain, takes another look around, and sees another light, fainter this time, farther up the road. He hoofs it up to the other light, slogging through weather up to his ass. The light is shining from another farmhouse. He pounds on the door, and an old farmer in a housecoat answers.

"So, Buddy says, 'hey, I'm stuck in the ditch. Can you pull me out with your tractor?'

"The farmer asks where he's stuck, and Buddy points down the road and says, 'about two miles down that way.'

"The farmer says, 'shit, son, that's a hell of a walk in this weather. How come you didn't stop at the neighbour's right nearby?'

"So, Buddy responds, 'well, to be honest, I did. But they were pretty fucked up.'

"The farmer says, 'fucked up? How so?'

"So Buddy says, 'I walked into the house, 'cause the door was wide open, and a woman is standing there, buck-naked, yanking on her tits. But that's not even all! There's a guy sitting in the corner, and he's pouring water over his head while he's cranking one out. See what I mean? Fucked up!'

"So, the old farmer starts to chuckle and tells the guy, 'no, no, you got it all wrong. They aren't fucked up. They're deaf and mute. She's telling him to go milk the cows, and he's saying *fuck you, it's raining*!"

Rossko howls with laughter, slapping his knee, and his enthusiasm wins us over.

It took a while to get there, but it was worth following him along.

"Well, boys, let's talk work. We've got this guy I'm going to send down with you. He's a little bit …" He pauses, carefully choosing his next words. "Twitchy."

Me and Wilton look at each other. Then I raise my eyebrow. "Twitchy?"

"Yeah, little twitchy," Rossko says. "I'm going to put him with you guys for a bit."

"Twitchy, Rossko?" I repeat.

"Hardly noticeable," he says with a smile. "It won't be too long. You two can handle it."

An hour later, I see exactly what Rossko meant by "twitchy" when a man walks towards me, jaw bouncing like he's chewing gum when there's nothing in his mouth. He's bone-white, almost see through, and I lose him in the glare of the sun, like a crack against a wall. The hair on his head has abandoned him, like I imagine so many other things in his life.

"Hey, um, yeah, um, how are you uh … yeah, um," he says.

My eyes widen. *Fucking Rossko.*

I analyze the guy, but he's freaking me out. I'm not scared of scary, I'm not frightened by tough, but I don't like unpredictable. It's not a thing you need in an already dangerous worksite.

"Fucking, um, ask how your, uh, day is, you, um, just ignore me? Uh, come on," he snarls.

"Oh. My day? Christ. Uh, it's good," I say.

"Hey, um, uh, yeah, I'm Brant." His eyes dart around his skull.

"Nice to meet you …"

"So, yeah, um … yeah, um, I guess we … uh, um, tomorrow?"

"Uh-huh. Fort St. John," I say.

I'm going to kill Rossko if this guy shivs me for a buck and a quarter. Goddamnit.

I go back to packing the truck for tomorrow,

telling myself it'll be all right. Crackheads work hard… when they're on crack … and they love to do the dirty, grimy work. The shit work.

Wilton glides up beside me and taps my shoulder. His eyes are so wide that they look about to pop out of his skull, eyebrows mangled in a confused, frightened look.

"You meet that fucking crackhead yet? What the fuck… I think he was chewing on his goddamn tongue."

"Yeah, this is bad. Terrible. Let's just pray it's for a few days, maybe a week. We've had worse," I say.

"Uh-huh. And when was that?" Wilton says.

I don't answer. I don't have an answer.

The next morning, we're running late, and it's a ten hour drive, so we're trying to make time. I'm commanding the water truck, Wilton is El Capitan of the picker truck, and poor Chiggins is getting acquainted with Twitch in the crew truck. Our goal is simple: keep the rubber on the road and between the lines, drive until we're done.

Three hours into the ride, I get a message from Chiggins.

This guy can't drive. He's almost killed us twenty times.

Take the wheel, then, I respond, thanking the

pipeline God's that I'm driving solo.

Wilton and I arrive at the hotel at almost three in the morning. We're asleep before the hotel door closes behind us.

Wilton's ringer jolts us awake less than an hour later.

"You better have a good fucking reas—What? What do you mean the tire fell off?"

I'm half-awake, Wilton's half-asleep. I can't tell if Wilton isn't understanding or if the explanation isn't making sense.

"Are you guys all right?" Wilton asks, swinging his legs out of bed. "Okay, well, that's good. What happened? No, tires don't just fall off. How's the trailer? Good. How far out of town are you? Good. See you tomorrow." Wilton hangs up.

"Tire just fell off the truck," he tells me.

"Tires don't just fall off trucks, Wilton," I reply.

"I know."

The next morning, Wilton and I, coffee in hand, stare at a truck missing its back right tire.

Wilton takes a long, loud sip. "Well, I'll be damned. Fucking tire fell off the truck."

We take the deck of cards and shuffle things around. Couple of the boys will jump in with us, while Chiggins and Twitch will get the truck towed back to an auto shop before taking another truck and hooking up our trailer and bringing it out to site. Simple.

We get another message from Chiggins.

This guy's fucked! We've been ready to go for an hour, but he is out shopping for clothes. Now he wants to have a shave and a shower.

Wilton calls Twitch immediately. "We gotta go, bud. We're late. There's no time for a shave and a shower. Like, what the fuck? We got work to do."

I can hear Twitch's voice vibrating through the phone.

Wilton looks at me with complete disbelief and turns on speakerphone. Twitch's voice booms through the parking lot, ranting about rotten marshmallows, sour apples, hard done by, no sleep, pity me, pity me, needs a shave and a shower, new clothes, working hard days.

Wilton and I start laughing at him into the speaker.

"Chiggins," I say. "Can you hear us?"

Small pause. "Yeah," Chiggins says through the phone.

"Leave the twitchy bitch."

Wilton hangs up the phone, order given and happily carried out.

Some Bad Juju

Things about Twitch start to make more sense when we find out about his past. I could see the man's energy, and it was black, and he carried it everywhere, spread it to everything he touched. This bad juju would follow behind him like a cloud.

Eventually Twitch was picked up, and now we watch him arrive at the job site.

He starts pulling out this giant tub, from which he pulls out another tub, and then a third tub of clothes from the truck. It's like a crackhead Russian doll. He starts hauling them up the camp stairs and gets ready to check-in. Another crackhead move.

I don't want the man dead. I don't want him to be mauled by a bear or unfortunately squashed beneath some falling action of a hoe bucket.

Things start happening around the site; strange incidents. It started with the tire falling off the truck, carried over to the first day when a hydraulic hose blew. Twitch started to help a guy with tool changes and somehow tempered steel tongs exploded like a crack pipe in his hands. Like I said, I don't want this man dead. I don't want cannibalistic tribes to shrink his head to the size of an apple and use it in a stew.

Our nerves are shot, boiling over at the events unfolding. A minor slowdown can have ripple effects that delay jobs for hours, which leads into days.

I watch from my drill as Twitch chews an invisible wad of gum. I can hear him in the back seat, smacking his lips. *Clip, clop, clip, clop*. The radio has to always be on to drown out the sound of his chewing. He scratches at ants beneath his skin and shouts small noises of protest against the bugs in his body.

I don't want this man dead. I don't want revolutionaries to disappear him into the voids of scientific experiments with chemical gases that weren't meant for the world of humanity.

We are driving back to camp in a loaner truck, since our crew truck is still in the shop. Twitch has his music playing; delta blues with sadness writhing through the notes. Whoever wrote these songs, Twitch could relate to their sadness. I like the music, and it nearly drowns out the sound of his incessant chewing. Something about the lyrics, sad songs about being beaten down by life, make me feel for Twitch.

"Your name's Brant, right bud?" I say.

"Brant, yeah."

"Brant, I got to ask … what's with all the body movements? You're burning calories just by existing back there."

"Yeah, um, I got, uh, ADHD. Yeah, really bad. Um, had it since I was a, uh, kid," he says.

"Okay, okay. That's some hardcore ADHD, man."

"Yeah, yeah. Yeah, part of life. I take, um, medication for it."

"Can you hook a brother up?"

"Sure, but it, ah, um, will … will … will put you on your ass." He chokes out a laugh.

"You play any sports when you were younger?"

"No, no sports. Music."

"Sweet, bud. Where'd you grow up?"

"I was raised by my, um, grandparents," he says.

I can't tell if this is difficult for him to speak about, or if he just finds it difficult to speak at all.

"I never knew, uh, my parents. Um, yeah, guess they were, uh, pretty into drugs and my, um, so my grandparents took me."

"Fuck, bro. At least they raised you," I say.

Twitch makes more sense now.

A shiver of sympathy shoots through my body. I feel for the guy.

"Yeah and, um, when ah, I was sixteen, I, um, tried to kill myself, but, ah, no luck," he says with a chuckle.

"Fuck, really? How was that?" I ask excitedly. I'm always interested when people try to end themselves, before the final moments, before they drift away to the light. Did they still fight to survive? Did they cling for life? Clearly, he found a reason to live and fought to stay alive.

He looks surprised about my excitement but continues. "Yeah, um, well, I was in a, um, sad time, and I, um, tried to hang myself from, uh, the rafter. I was struggling. Um, fighting. Then I went out. Um, I guess the rope snapped, and I fell, and ah, made a thump, um. My grampa found me, um, and got me to the happy hospital."

"Fuck, man. Why would you hang yourself?" Twitch goes to answer, but my excitement steamrolls over him. "That's so boring! I would jump from a plane, fly through the air, then *splat*! Or a fuck-ton of drugs. See what the human body can take, you know?"

"Oh, yeah. Um, when I worked in, ah, Cali, I got

pretty, um, hooked on the drugs," he says. "Crack, uh, meth, ah, angel dust, and ah, uppers. Lots of uppers."

"You don't say."

The topic continues about how he was raised by his grandparents. His original parents left him for a glass pipe, and the grandparents took over. Abandoned at birth for a couple pieces of gear. *Fuck.* The pieces were coming together. Twitch had a sad life. All he knows is how to break, blow up, and abandon.

The one handy thing about a crackhead is they love to do the dirty work. Another hose blew, and the rigs blood is mixed with mud and snow. Twitch is playing in the mud, and this ghostly white motherfucker is caked in slop, smiling away like he truly belongs in the dirt. He's crawling through the trenches before standing and walking right into the passenger seat of the truck in all his muddy glory and puts his gloves on the dash and steering wheel, seat smeared with globs of glop. He goes into one of his tubs of clothes and pulls out new coveralls.

Wilton goes off, spitting fire and nails. "What the fuck are you doing, man?"

"Yeah … Uh, what?" Twitch says.

"Are you fucking kidding me? What is going

through your cracked-out fucking skull?"

"What, um … I was, ah, working. I, ah, need to, ah, change," he stammers.

"Why with the mud, man? Fuck! It's everywhere!" Wilton is so taken back that he doesn't know what to do. "Change and clean this shit up!"

Wilton looks at us, his eyes roll. He slams the door of the truck behind him.

I don't want the man to die. I don't want him to choke on pieces of bacon and drown in a bubble bath.

One day, we let Twitch drive back to camp. A blind man with one arm driving a stick shift could have done a better job than this guy. Chiggins was right; he's a terrible driver. He sped into glare, hit black ice, and we hit the ditch. Rattled the whole crew, put the fear of God into us for a second and a half. Twitch's sadistic melodies were playing, setting the mood like a movie soundtrack.

Approaching the highway, there's a clearly marked sign, red with white letters blazing through the dark saying *STOP*, and Twitch puts the pedal to the floor.

Panic. Is he trying to commit suicide or mass murder?

We're all screaming wild profanity until Twitch

mashes on the brakes. The wheels lock up and the vehicle starts spinning. Twitch is staring straight ahead, hands off the wheel.

We come to a stop, bed-end forward, touching the sign. I can hear our hearts beat in our chest, in my ears. A semi passes, and the bright of its high beams flash our lives before our eyes.

I don't want this man to die. I don't want a flock of ostriches to maul him to death while lightning bolts shoot electricity through his veins … Why am I thinking this?

Camp

The simplicity of camp is great. Look at it like this: there's a bunch of wild animals working, connecting pipe, drilling holes, coding joints, all this in the deep bush, sweating out liquor, covered in muck, burping, farting, all smelly bastards. The job site might be anywhere from an hour to three hours from the nearest town, so they build these camps closer to the job site to feed and lodge us all, keep us from polluting the town with our existence.

The camps are built a bit like Lego's. The pieces are all long rectangles, organized into a slightly altered design on the board. Every camp is a little different; however, the interiors are always the same.

Walk into the front entrance and see a lady at a desk. Most often, she is fat and unattractive, and that's why she works at a camp away from civilization. It's just one of those sad truths. Next, you sign in.

"I'm Joe Blow. Here to work."

She'll reply with a stern look in her eye. "You're

on the list. You know the rules?"

Then you answer with a serious face, "Yes. This is a dry camp. No booze or drugs."

She'll hand you a key.

Next, you walk down the corridor, quickly check the games room to see the same standards: pool table, ping pong table, foosball table, surrounding a couple of semen-soaked seats and a big screen TV with satellite. Beside the game room, there's a gym. This one is actually a good one, with free weights, an actual bench press, and treadmills for some reason.

Finally, the personal lodgings, Chateau du Dirtbag. It's either a single or a double bed in the middle of a room that is four paces from wall to wall.

I throw my bag near the fridge. Everyone's different, but I choose to get all my toiletries out and organized in the bathroom. Then I head off to the feeding trough.

Some camps have good cooks; some unlucky ones have bad cooks. Hell, does it ever make a difference after a hard day at work whether you have a good meal or a bowl of slop. Luckily, we have good cooks here.

Those in line for food are all different shapes and sizes. It's amazing how different a guy will look on the pipeline in his coveralls than in his sweatpants and a T-shirt in the dining room. A man covered in mud-stained coveralls, hard hat, toque, and safety glasses loses proportion.

There are beanpoles with the bald spot, the lazy lards lumbering, big-boned boys, and two plate eaters. There are guys like me who like to hit the gym too much. There are also the plump fellows in their fifties who basically live at camp. And, of course, there are other boys living the pipeline dream who wouldn't look out of place in an accountant office.

We scarf down our food at the same seats with the same people, like we're in high school.

Next, because it's a dry camp (feel free to laugh), I head outside to the parking lot and hop in the truck that constitutes the pipeline social club. All these trucks full of boys, smoking darts, joints, and slinging back beers. Whatever will take the edge off the day and help forget what problems are waiting at home.

Where do we get the beer from in a dry camp, you ask? Oh, that's simple. Before we leave the next large town, we hit the nearest Costco and buy in bulk. We drink these small towns dry.

After the mind is numb, you're ready for bed. And, just like a child, you sneak past Mom (security), and slip into your four-by-four room for a shower. Dive into bed, throw on the television, grab your phone, firmly grab your cock, and choke it while you watch a hardcore film of a girl making a banana disappear. Release bliss into a napkin and slowly drift away to sleep …

A Fight Over A Dollar

Global warming, what a beautiful thing! For the last hundred years, we humans have drilled holes in this poor planet and slowly sucked its blood out, just to burn it so we can go faster, destroy you quicker, and now it's raining in December.

The rain hits the road and turns instantly to glare ice. The job gets shut down because of the driving conditions. There are twisted bodies of semi-trucks in ditches, jacked-up Ford's rolled over, spreading glass over the highway. Death is around each hairpin, but we have a day off, and we are thirsty, hungry for adventure. Death will not stop us. It only adds to the fun.

We get to the camp with a jump in our step.

"Don't worry, Chiggins; I got a guy in Fort St. John who can hook us up," Wilton says. "Give me a quick sec."

"Hey, um, ah, guys, can you a get me some, um, blow, too?" Twitch says.

"Sure, man." Wilton laughs. "Let's head up to

Station 101 and get supplies—beer, fried chicken, all the works."

The grease-soaked skin and the tasty crunch of gas station steroid chicken can't be beat. Brucie sees us taking off and shoots Wilton a message to grab him a bottle of rum and a twelve-pack of beer. Brucie's an alcoholic and a damn successful one at that. He was a deep-sea welder for twenty years and a heroin addict for five. He's died twice and cursed out the people who brought him back from the grave each time for shocking the heroin right out of him. He's been beaten down by self-induced bad luck, but he keeps ticking.

As we slip and slide along the highway towards the camp, our back seat rattles around with two-six of rum, a two-six of fireball, seventy-two ice-cold beers, one quiet boy, and one twitchy cracker.

Brucie is added to the truck party and instantly starts slamming beers. The boys mash, crush, and start dividing up the blow.

"Before you jam that wonderful white wacky powder up your noses, we need an epic song drop," I stop them from hitting the lines.

I queue up "Something in the Air Tonight" by Phil Collins and instruct everyone to wait until that gorgeous drum fill caps. The waiting is killing them; I can see it. They keep thinking it's going to drop early. Anticipation has them jumping like Twitch. They want those lines so badly. *Sniff me*, the lines beg them, but Phil teases him with a fake bass drop.

Wait for it ... Wait for it ...

Bum-bum, bum-bum, do-do, bam-bam, puff-puff. "*I can feel it coming in the air tonight ...*"

Oh Lord, hungry noses devour the lines, and they throw their head back, feeling whole again.

Booze cruising on black ice. What a riot. We speed down twisting, winding lease roads until we reach a large empty field. A bunch of idiots of all ages in a field below cold, spitting rain, cranking tunes and drinking beers. This is how I want my life to be. You can't control the flow of nature or plan every aspect of life, every second of every day on a schedule. We are uncontrollable. We flow like a river and arrive wherever the current takes us. We should be unpredictable.

"Casino!" Wilton shouts. He turns to us like a carnival barker, persuading the crowd. "We got the day off tomorrow, boys. Think of all the money we could make. Think of all the fun we could have."

"Fort St. John is, like, an hour away," I answer

like a bitch.

"We need to have some real fun," he says, eyes judging me.

What was I thinking, arguing with a devil who won't ever quit! "Fort St. John, baby, here we come."

The front three seats are a riot—Wilton driving, Chiggins in the passenger seat, gripping the holy shit handle, and I'm in the middle, chopping lines up on our work binder. The back is a different story.

Brucie is so drunk that he's passed out, mouth wide open, staring upwards like a chicken looking to drown in a storm. Twitch doesn't look right. Perhaps the past that he's been running from is catching up with him. I can see his demons. This is going to be a night, I can feel it.

*

Well, fuck. Casino sucks. It's eleven at night and none of the tables are open, just a big room with tall mirrors filled with retirees playing VLT.

Wilton exits the casino and vomits along the side of the building, either because he's shitfaced or because he handles disappointment poorly. It's tough to tell. If he's shitfaced, well, he did a pretty good job of driving through terrible road conditions. Luckily for us, the strippers are open.

We saunter in like we own the place. Well, except for Brucie. He's still in the truck, dreaming of heroin, beer bottles, rum, and lines. Twitch is one sketchy-looking cracker with his skinhead aesthetic, constantly touching his face, checking for bugs, speaking in crackhead stutters.

"Hey, yup, ah, um, can I, ah, have one hundred dollars of loonies, um, ah, please, um, please, ah, please," he asks the bartender.

I scan the scenery. Two pool tables below neon lights that wrap around the walls, dust on the fake diamond chandeliers, broken glass in the corner. The bathrooms are marked with a sign over each saying 'Dicks' and 'Pussies'. And front and centre in perv row sit Chiggins, Wilton, and Twitch, and they're saving a spot for me.

Chiggins is broke like Humpty Dumpty, but somehow, the strippers flock to him and make conversation like they aren't trying to sell anything.

The boy has a gift. There are four men on perv row, but only one of us is sitting with a stack of a hundred loonies in front of him, darting his eyes in all directions while his jaw makes circle motions.

The show starts. Bombshell on the stage with all her fake, bulging parts. She breathes fire like a beautiful dragon. I guess her tits aren't noticeable enough, so she lights those on fire.

My cock is rock-hard as she bounces that Jell-O ass in front of me and lights that on fire, too. My senses are overloaded. I throw all my loonies in one big load, spraying them on her tits and face.

"Clean yourself off," I say.

"You, too," she says, giving me a wink and sliding her hand across my crotch.

Chiggins new friend is on next, and she directs the show towards Chiggins. Of course, he doesn't have a dollar to his name, let alone one to donate. It's painful. I can't stand to see Chiggins not throw something at her pussy, but I have, unfortunately, unloaded all my money in one go, only a few minutes prior.

Thankfully, Twitch has the loonie Tower of Pisa in front of him, so I slide one off the top and push it towards Chiggins to please his gal. Chiggins throws and misses her vagina, and Twitch explodes.

He's so close to my face that I can smell his breakfast, feel the bugs under his skin touching mine. I back up and clench my fists.

Wilton and Chiggins separate us.

I'm calm. I like to fight, and this mid-forties, bag of skin and sinew is riddled with weakness.

"Don't you, um, touch my fucking money!" he yells.

"It's a dollar," I say, rolling my eyes.

"It's, ah, my money, ah, don't fucking touch it."

"Again, it's a dollar. Don't you think you're overreacting?"

"Fuck you!" he screams. "That's, ah, my fucking, ah, money, um, and I don't, ah, want that little, ah, shit to have any."

Chiggins mock-frowns. I guess Twitch doesn't like him.

The boys bring Twitch outside where he's still shaking, vibrating even more so than usual. What a scene of absolutely nothing.

The crackhead has a point, though. It's his money. The money he worked hard for, money he bleeds for, money he sweats for. I didn't have the right to give Chiggins another man's cash, even if it was only a dollar.

I go to the ATM and pull out five crisp twenty-dollar bills with a flick of a card. I then go to the bartender, who has worked here since she was in her mother's womb (family trade, I suppose). Her face is almost tanned from the glow of the neon lights, and the lines on her face run deep with cigarettes and booze. I order a round, collect my change, and then walk outside into the brisk northern air. They're still trying to calm Twitch down.

"Brant, I apologize. I should not have taken your money. That's my bad. Here, I got a dollar I owe you. Again, you work hard for your money, and I had no right to take it."

"Fuck you. Ah, I don't want, um, you're ah fucking, ah, money," he says, scratching his chin and cheeks.

"Man, take your dollar and let's be done with this. I bought us all a round."

"It's the, um, the principle," he says.

"I said I was sorry, and that's all you're going to get. I have your dollar. I understand you work hard. You're right that I had no right to take it. You want two dollars? I'll make this worth your time." I stroll up to Twitch and put the dollar in his pocket. This infuriates him.

He pulls it from his pocket and throws it across the parking lot.

The boys try to explain to him that we are square,

that it's not worth continuing over.

"Brant, he said he was sorry. Like, it's over," Wilton explains. "He gave you your dollar back."

"I threw it at the girl. I'll give you a twenty when I get paid next," Chiggins offers.

"Fuck, ah, you guys. It's, um, the fucking, ah, principle. He doesn't, uh, yeah, mean it. He's just, um, saying sorry. But, ah, he's taking the piss, ah, like it's a fucking joke. It's, ah, my money. Um, it's not a joke. I'm, ah, not a joke!"

I've had enough of this. The boys are wasting their energy on a crackhead tantrum. I won't let one more breath be wasted.

"Leave this fucker out here if all he wants is attention. Let him ramble to the parking lot. Brant, stay out here alone. We're going," I say with intent and anger.

We sit at our seats, and the boys are happy to be greeted with fresh beer. Twitch is going off in a tidal wave of madness, winning arguments to no one beside us. We don't look at him or pay any attention, but his negative energy is driving away the girls, so we leave him in perv row and head to the other side of the bar. When he sees us leave, he turns even more into himself.

"It's the, um, fucking, ah, principle!" he yells, winning whatever battle is raging in his head. Then he grabs his stack of loonies and throws them across the bar, and after a moment of stunned silence, the strippers begin collecting them off the stained carpet.

I can see other clientele looking at the scene and putting two and two together that he's with us, but we're trying to talk to the girls, drink our beers, and relax. I can see them staring as we drop money to paper over Twitch's tantrum. I can see them staring at us, wishing they were us. Yeah, everyone wishes they were us …

Wilton

We pick up Brant the following morning. Turns out he wound up in the hospital sometime during the previous night. He was acting like such an undesirable human being that they kicked him out of where they keep the sick and afflicted.

Our crew truck is fixed from the shop, so we have Chiggins follow us with it. We leave Brant the loaner truck, and he drives to camp alone. In many ways, he's been driving the same lonely road his whole life.

"What a low end," Wilton says.

"What was that all about last night? What a lunatic," I say.

"There's a reason why society hates crackheads," Chiggins mumbles. "They're unpredictable."

"Low end. Fucking low end. Like, you try to be good to a guy, and he blows his top over a dollar, a fucking loonie!" rants Wilton.

"Once a crackhead, always a crackhead," I say.

The next day, we're at the safety meeting. The

boys stand together proud, chest out and confident, looking down at the snicklefritz and outcasts. After the meeting, Chiggins and I take our positions in the crew truck while Wilton talks to the spread boss, likely about the rumour that Twitch ended up in hospital.

Turns out that, if you speak of the devil, the devil shows, because lo and behold, Twitch is walking towards our truck.

I lock the doors.

"Chiggins, don't you touch those locks."

Twitch claws and bangs on the door, and Chiggins, the coward, relents.

You could cut the silence with a knife. English needs a word more powerful than 'awkward' for situations like these.

Wilton arrives and puts the truck in drive.

"We gotta answer for that fucking bum," Wilton rants. "I mean, what a goddamn waste. You try to be nice to a guy, take him out, and end up answering for him over a buck."

Wilton is obviously oblivious that Twitch is in the seat directly behind him.

Chiggins and I stay silent. We don't want to interrupt this well-needed roast.

"Once a crackhead, always a crackhead. People don't forget. And we don't forgive out here, not if you want to be one of the boys," he says. He's on a roll.

Twitch's face is hanging low, jaw hanging loose. He remains silent, and so do we while Wilton continues.

"He lost his shit. All of it, all his shit, over four quarters. Like, what a fucking loser. That's why no one in the company likes the guy. Did you see him in the safety meeting, standing all alone? No one even wants to brush shoulders with him. Then he texts me that he's in the hospital, and we gotta pick his ass up? You go out for a few beers and end up in the hospital? What a fucking low end."

I tug on Wilton's shirt. I look him right in the eyes, a smirk rising on my face. "Bro, he's right in the truck."

He turns around in disbelief, shocked that he's been talking shit behind someone's back while he's right behind his back.

Wilton is a deer in the headlights, stuck in the headlights, halfway between two options:

1. Roll over and die (apologize).

2. Charge head-on and hope the car swerves (God hates a coward).

"What the fuck are you doing here?" Wilton howls. "You're gone! You ain't one of the boys. You're done."

"Are you, um, serious?" Twitch replies, his voice coming out softly from some faraway, abandoned place.

"Yeah, I'm serious, you cunt. What was last night? I think you owe all of us an apology. And now that we're on the topic … a fucking dollar? Are you serious?" Wilton reaches into the cupholder and extracts a fistful of change. "You hard up? I've got you, bro." He throws the change at Twitch's chest.

"You guys, ah, left me there," Twitch says.

"We told you we were leaving!" I bark. "But you're so fucked in the head that you just pouted and made a scene like a baby-back bitch."

"Is that how you hang out with the boys?" Wilton says. "Have a few beers and turn into a complete monster?"

"You, um, guys are serious?" Twitch says.

"We don't want you around. You're shit," Wilton says. He turns and looks Twitch dead in the eye. "Fuck off."

Twitch scurries from the truck, head hung low.

We watch him kick dust up behind him from the

vehicle.

"What the fuck, guys?" Wilton yells, still hopped-up on anger. "No one told me he was in the truck!"

We explode in laughter.

A sacrifice to the oilfield God's, and Wilton didn't even know he was drawing blood.

"I ripped into the guy for, like, five minutes, and you didn't stop me?"

My lips curve into a sinister smile. "I didn't want to interrupt."

"I hate talking behind someone's back," he says before laughing. "Well, I hate getting caught."

I'm howling with laughter to the point that tears are coming from my eyes. Chiggins is doubled over, clutching his chest and forgetting to breathe.

Either you're one of us, a comrade, or we cleave you from the herd and leave you an outcast. Twitch was sacrificed on the altar of the pipeline. May our prayers be heard by the God's.

Black Coffee and a Birthday

I've worked a lot of jobs with a lot of foremen, and it's typically the same: meet the new boss, who is the same as the old boss. But this job is proving that change is possible, change can even be good.

The foreman for this job is a guy by the name of Tyron Padoso. Tyron is covered in tattoos, everywhere but his face. His body looks like a downtown mural—colourful and eclectic. He looks terrifying until you talk to him and his soft features light up when he smiles, and it becomes clear he's a softie. Perfect quality in a boss—stern, tough, ready to work, but willing to forgive a bit of slacking if you can crack a good enough joke.

We arrive on site, and Tyron stares us down.

Wilton shines a charming smile.

"You boys got some black coffee?" Tyron barks.

We nod.

He hands us a bottle of Crown Royal and a bottle of Bailey's. "Black coffee is for retirees. Why don't you spice it up?"

Now this is a man who knows how to make an introduction.

Am I the only one who routinely forgets how old I am? Birthdays mean nothing, not even a number, but mine's in a few days. Wilton is more excited than I am.

When you work grinding gravel roads and losing track of days on the pipeline, birthdays don't matter. Another day, and every day might as well be a Monday. But, to my best bud, it's time for payback.

Wilton has missed my birthday for the past five years and is seeking redemption for all the birthday parties I threw for him. I'm lucky to have such a friend. What is friendship if not destroying one another at the right times?

"Eh, bud, your birthday is coming up. You pumped?" I go to answer, but he cuts me off. He's vibrating. "I'm so pumped. Twenty-nine years old."

"It's not thirty," I say. "They don't make cards for twenty-nine."

"I got all the boys coming out—all of them. Table reserved at the Brewhouse. Everything will be taken care of. There's going to be like twenty guys. Absolute party fest. Then after …" Wilton smiles. I know what he's going to say. "Casino!"

"Perfect," I reply.

Wilton's excitement is a contagious growth. He can get anyone excited for anything. It's the plague of the pipeline. Well, herpes. Syphilis. Chlamydia … It's one of the plagues of the pipeline. I try to hide my excitement, because things can fall apart quickly. Long day, work late, an emergency happens and, suddenly, the best laid plans are undone. Best not to have any expectations.

We get a call from Tyron as we drive to work that morning.

"Boys, make sure you bring lots of black coffee."

"What's going on?"

He's vague. "We got a little problem."

"What! What is it?"

"Just a little problem where we got to set up," Tyron explains. "Engineers changing their minds again."

"Well, what do you mean?" Wilton asks with disdain. "We need room to make our cuts. If we're too close, it'll fuck us."

"Just bring some black coffee, and we'll talk about it."

We pull up next to the truck with all the boys around. Again, Tyron passes us a bottle of Crown

Royal and one of Bailey's. He basically has a bar in his truck; roughly eight varieties at any given time. We pass him a bottle of Irish whiskey from our truck. Nothing is said until all the boys have improved their mornings.

"Boys," Tyron finally says, "the fucking engineers changed their plans. I need you to set up closer and make a hard cut."

"Tyron, our rods can only bend so much," I say.

"Just drive some of that black coffee in ya and make the cut."

"We'll see, but I'm going on record as being against this," Wilton says, sipping his now premium coffee.

We pause for a bit, enjoying our morning. Order given, order received. Even if we disagree, we're still going to do what we were told.

"So," Tyron starts, "me and my friend are going to the AVN awards in a couple weeks. Vegas."

"What's that?"

"The porn awards. The guy I'm going with is a pussy-slaying machine. Tattooed all over his body, like forty grand worth of ink. And this guy smashes bin. This one time, we were on a flight to Vancouver, and he convinced the girl next to him to suck his dick.

That flight is like an hour and a half from wings to wheel." He laughs.

"The porn awards?" I laugh. "Shit, ain't you married?"

"Yeah, but you know," he says. "She gets all her nice things, and I get to go on a trip to Vegas once in a while."

"Does she know you're a dog?"

"Of course not! Guy needs to release the stress, though. Especially managing you idiots. Shit, I fucked the medic up the road just yesterday after work. Had a little black coffee with her, too."

This gets us howling. "Fuck outta here. You didn't."

"No lie. But don't say a word. Don't need the boys spreading that shit around."

"Okay, we'll see what we can do with the hard cut," Wilton says. "But that secret is going to cost you a few extra *hey, how you doings*."

*

Wilton's ecstatic. It's my birthday. This will be the first time in five years I celebrate properly with friends and had a real party. I'm excited, to say the least. I've got the anxious excitement where every small obstacle feels like an enormous burden stopping me from my goal of getting obliterated. I can't eat, too in the drinking zone.

Chiggins is pumped, too. He actually has money and is talking about all the beer he's going to bring upon us. He's dreaming big, beautiful dreams about tonight.

All day on the radio, I hear Wilton infecting people, convincing them to forget about tomorrow, charming them to live for the now, or outright bribing them to not just join the party, but to not stop until we kill the night. The day is spacetime, the finish is seconds after the start. There's a jump in all our steps. The gas pedal has a mind of its own and floors itself the whole way home. We're all flanged up and ready to go.

We arrive at the Brewhouse. All the dirtbags and lunatics shine up like new pennies. It's a shame that you can polish a turd and wrap it in glitter, but it won't disguise the inside for too long. Everyone is here: the site boys, the shop boys, welders, hoe excavators, even Rossko. We're all peacocking to hopefully excite some women. There are two beautiful bartenders who can't leave our group alone.

Wilton orders a "time of the month." It's whipped cream, dyed red, and under all that crap is a shot of Jager. One of the bartenders, Bree, happily puts it in front of me as I dive in (look, Ma, no hands!) to find my reward. I emerge a mess.

Bree licks the whipped cream from my finger, and I have a half-chub now.

The bar is a carnival. Beers crash and bang, spilling over the table. The boys scamper like ants to and from the bathroom to feed their noses. Shots are bought en masse. Tyron couldn't make it, but he ordered a round of shots over the phone.

Wilton goes to his car and returns with an ice cream cake, and the bar erupts with "*happy birthday, dear dirtbag*," and every conversation that isn't ours is drowned out.

The boys can't eat much on account of the drugs, so we start giving all the barflies pieces of cake, singing to them a bit of "Happy Birthday" as we do.

I fork-feed Bree some from my plate and get her number. I tell her we're gambling our life away and that she should join us. Then we clear the tab, and all us mad men head for glory.

Roulette is our poison. You can't recreate the feeling when the ball is spinning around. Bet on black. No, bet on red. No, black! Wilton bets on random numbers, claiming he has a system. The ball spins around. Wilton bets hard. Big wins, big losses. One of his numbers hit, and we strike gold. Wilton's up four-hundred dollars. I'm slacking behind with three-hundred. The ball goes around and around; we buy round after round. Bree and her friend arrive.

"Pick a number, baby, any number. As long as it's a winner," I say as I hug her close.

We go on a pilgrimage to find the right VLT. We sit in the squeaky, swivel chairs usually reserved for the daytime elderly, and I can feel the tension between us. I pull her face close and our lips meet, then our tongues follow. There's silence when I pull away.

She takes the money and slides it into the machine. The machine takes our money away too quickly. Barely gives us a chance to dream.

Chiggins messages me that he's in the truck, so I head over there, sending Bree back to the boys so I look better by comparison.

"Buddy, what's going on?" Chiggins says. He's sprawled over the back seat, shoes off, looking quite comfortable.

"What's shaking?" I ask. "Who's this?"

"Hi, I'm Dani," the woman greets me. She has long legs and an attractive face.

I'm trying to place where she's from, but I can't put my finger on it. She looks similar to girls I would see during my travels in Asia, but my head's too foggy to remember specifics.

"You want a cigarette?" Chiggins says.

"I'm drinking, so I'm smoking."

Chiggins passes me a dart and starts busting out lines for him and Dani. *That a boy, Chiggins, try to buy that pussy*.

Do drugs and liquor make people wilder, or does it just give them an excuse to be? I'm no scientist, so I don't know, but after a few party favours, women are a lot wilder, much more sexual, even if they have boyfriends or husbands. You just need some patience. Time is your friend. So Chiggins continues to pass the time by racking up lines.

Dani holds one nostril shut, snorts with the other. She then leans her head back and swallows the drip.

"You know I'm a lesbian," she says.

"Good for you," I say. "So, you're off the pole for good? Save the cowboy and ride a horse kind of thing?"

"No, but women just know how to please a

woman," she says with seductive eyes.

"Don't like the rough, rugged pounding, thrusting inside of you?"

"I'm indifferent," she says. "But I was thinking about getting back with a man. You know, try it out again."

"Oh, really?" Chiggins says, perking up. The dog in him hears the rustle in the bushes.

As we walk back by the smoke pit, we see one of the pipeline boys from North Sask. He waves at us cheerfully with his pants around his ankles, prominently thrusting his cock down some woman's throat, right in the open. What a bizarre sight, yet somehow, it doesn't register as crazy. We just keep walking past their choking passion, and slowly, it fades away, like it was a shared mirage. As I write this, I wonder whether it really even happened.

It's nearly two in the morning, and in a few short hours, there will be an army of hungover men on site, not even far enough removed from the night to be sweating out the poisons yet. Most won't have stopped. But we will all show. We will all suffer, but we will suffer together, same as we always do. That's what brotherhood means.

53 Days Without A Day Off

It's day fifty-three in a row without a break, and I'm wearing anchors. My thoughts are moving through molasses. Depression has slowly crept in, day after day. At first, I felt it like something out of the corner of my eye. Then I let it linger there, watched without moving as it came closer, welcomed it to envelop me, embraced the burden as it turned my mind black.

Fifty-three days isn't even my record. Eighty-seven days is my record, although it isn't recommended. Depression came in all the same then, too. This one is hitting hard. A new site, thoughts of Nami still rolling around my mind and settling at the bottom, weighing me down.

I can't remember the last night I went to bed sober. I scoured the streets of Lloydminster for one of those movie punks selling weed and found one in a back alley with his hood up, trying to cover his face. Classic.

Over the last week, the depression has kicked

into overdrive, like a distance runner sprinting to the finish line at the end of the race. It started with a few beers over dinner, then it was skipping appetizers and getting the drink deal of the day. Waking up all dried out like leather pulled tight, not shitting solid. I tried to have the weed smooth the cracks of my character, maybe lull me to sleep. When that didn't work, I started lying in bed with a case of beer, feeding them into my belly until I passed out.

Each day, I got worse, deeper into myself, yearning for a day off, yet too stubborn to admit I couldn't keep this train moving. I can't show weakness. The boys need me. What's one more day?

The reality is that I'm just a greedy little fuck. The dark void inside started to retaliate. I'm sick of my desperate need for more.

Bad thoughts start emerging from the fog, the kind that can inspire or end a man.

Would the last man on earth feel free, or somehow suffocated by all the space, the totality of what wasn't there?

I ate dinner at Boston Pizza with a certain disdain. I swear I've eaten everything on this bloody menu in my time on the pipelines. Every oil town has a Boston Pizza, and it's the same place in each town.

I'm sitting alone in a booth made for a family, hating all the smiling people, laughing at me while they're enjoying life, being loved. I've turned sour. I hate the people because I hate myself. I hate the Madhatter who biked across Asia, the lunatic who bought land in Ecuador, the dreamer who hitchhiked the coasts of Australia and jumped off a cliff in Brazil. I hate what I've become—another pipeliner too greedy to ever leave alone the cash, the work, the beer, the day after day after day of aging with nothing to show but money. Would I even know if I was just a robot? Eat, sleep, wake, work, repeat.

More beer, now a high ball. Ah yes, the liquor is filling the void. I feel better. I see clearer. The cure to my problems is alcohol. How lovely. How fitting.

I leave Boston Pizza with a nice buzz at 7:39 p.m. I don't know why I remember that. Sometimes my brain just decides, *You'll remember this forever*. I don't remember what Nami's voice sounds like, not really, but I left Boston Pizza with a nice buzz at exactly 7:39 p.m.

I hop in the truck and drive to the liquor store to pick up a casual twenty-four pack for the night. In the hotel room, I start putting them back, one after another, breaking stride only to piss. The cracks start to fill in, and I feel better.

I want to be classy, so I play Bach through my laptop speakers and let my thoughts drift away while I think of nothing more than drinking and pissing. I'm quite drunk, but the void isn't filled yet. It almost seems like there's a leak in it.

I shake the case of beer and feel the cans shuffle around the empty space that I've created. It makes me sad in a deep, wondrous way. I'm losing control of the buzz, letting the loneliness back in; opened the door for the depression that was waiting patiently.

The void, unimaginable; the loneliness, intolerable; the depression, immense. I start talking shit to myself, as though I'm right in front of me, all my disgusting failure stacked high. I silently scream at myself, wanting to yell but can't bring myself to make a noise.

Slowly, I start peeling off my clothes, muttering to myself, *fuck it, fuck it all to hell.* I grab another beer, nude.

I am nothing. I am a black hole, a self-centered fuck. I am shit. I am absolutely right. The world doesn't need me. I'm worthless.

If we are doing this, then we're going to do it right. Maybe I could jump off the tallest building, be free as a bird for a little while. I could steal a Ferrari and try to jump the Grand Canyon. That could be fun. Or I could strap myself to a wrecking ball and wait for it to hit the building. As I think and think, more ideas come, more poetic ways to do myself in.

I stumble into the tub and turn on the shower. It feels like needles.

I would end it gloriously, like a Greek tragedy. I wouldn't hurt anyone like the cowards who kill their wives first, or bomb innocent people, or shoot up schools and festivals. I will do better than that. First, I will quit my job. I won't die a pipeliner.

In a single night, I'll plant ten-thousand roses on the hills of Edmonton's river valley. No one will know how or why, only that they woke up to ten-thousand roses, and every time they walk through the valley, the smell engulfs them, and they see rolling hills of red and white and purple. Roses with the biggest thorns, so when people come close, I can inflict a little pain. You can't have pleasure without pain. Then I'll take all the money I've made selling my soul to oil and gas and I'll burn it. Not for anyone else, only for me. I need to see it burn.

I have some quality organs I'll give away, but not my heart. That's for me. I'll carve it out of my chest with a knife, pull it out with my hand, and break it. I'll give everything inside of me to someone who needs it. My liver to a drunk, my kidneys to the sick, my eyes to a blind man. My tongue will go to a mute, my lungs to a smoker, my skin to those afflicted with burns. My brain is so dark and mad that I'll give it to science to dissect until they've found all they can find, and then they will throw me in the trash, but my heart is mine. The rest is yours to take.

As I plan my remaining days in the tub, under the hot stream of the shower, I slowly drift to sleep in my drunken stupor.

I wake several hours later, cleansed. Tomorrow, my brain will hurt, and my body will ache, but I will feel something.

Finally, A Day Off

Destruction is good for the soul. Burn the candle at both ends until enough wax melts to build another candle. After over fifty straight days of fourteen hour shifts, there's only one thing on my mind—getting completely, utterly, and devastatingly obliterated. I mean wrecked. Not some cute euphemism that to most mean a morning Tylenol and a skipped breakfast. I'm talking *fucked up*.

I've been away so long that the drive back fills me with anxiety, like there's a balloon expanding inside my chest. I'm a child waiting on Christmas Eve, knowing that Santa is going to stuff his fat ass down the chimney, bearing gifts. Except, instead of gift cards and socks, it'll be rum and whiskey, and I'll be one holly jolly son of a bitch.

I arrive home for the first time in months, and I'm both relieved that nothing has changed and a bit annoyed.

I'm not the only person who lives here. Old man Richard lives here, yet the front gate is as squeaky as the day I left it, and the only modification is the Folgers can in front of the back door has accumulated more cigarette butts. They stick out from the sand like headstones commemorating five minutes of Richard at a time.

Richard is a good man. Sixty-four years old, his fingers are stained yellow from cigarettes and his face is made of hard lines. He wears reading glasses atop a thick brown moustache and his throat is lined with gravel, so his voice carries a wear-and-tear of exhaustion commonly confused for wisdom.

Something about age makes a person complain. Maybe their body hurts, and they feel a bit out of step with the changing times and are compelled to rally back. But Richard is positive he only complains because the youth are shit. Time and time again over beers, I've mentioned how his generation ripped a hole in the ozone layer, cut down all the trees, and are still the limp dick bureaucrats making the rules, whereas my generation is just making peace with the bust hand we've been dealt.

Richard loves to complain, and I don't have the energy to fight inside my own home.

Richard is also well-versed on a number of subjects and knows exactly where the problems in our world stem from.

On politics: "It's because of you kids that, that fag Trudeau got elected. Because all"—he waves his hand at me and, I assume, my peers as he sparks a joint—"you kids don't understand the world, and voted that hippy teacher in." Richard also doesn't vote.

On technology: "Everywhere I go, kids always on your smart phones. Don't know now how to communicate! You know what I think? Do ya? They should be called dumb phones. Because that's what you kids are—dumb."

On warm weather: "Oh, it's too damn hot. You can't be working in this weather."

On mild weather: "Oh, it's too damn cold. You can't be working in this weather."

On Edmonton roads: "The roads are too muddy. I just cleaned my car, and it's dirty before I even get back to the house. Why don't we clean up this town?"

Yeah, he's a typical old man, but he's got a beauty to him.

The two things that old man Richard loves is hockey and his granddaughters. Those are the only topics he talks about with any revere. Those are the two things that give him a reason to live. Without those, he would be in the grave. Worm food.

The old bag of mouldy socks will not shut up about how bad the Oilers are. He's a Leaf's fan, which makes him a special kind of asshole in Alberta. Hockey is what makes the winter tolerable at times. The only sport that really unites the nation. A true gladiator sport. Mainly, because it's the only major sport where, when you take a cheap shot, instead of complaining to the referee, you throw the gloves off and smack your fists into another man until you both fall down or the zebras break it up. Richard, that old crab apple, as soon as I walk in the door, starts off.

"Did you see the Oilers play? It was pathetic. Absolutely pathetic. They're the worst team in the league!"

"Thank you, Richard. I know," I say, giving him the middle finger.

"You got to do something; you guys are falling apart."

"Oh, *I've* got to do something, eh? Me?"

"Call the GM and tell him they've got to shake things up, boy," he says. "You guys are terrible, just awful."

It's barely even begun, and I've had enough.

I head to rinse off with a trusty shower beer and

start fueling my mind into overdrive. When a man's roots are taken over with stubborn depression, the only cure is obliteration. Only truth will be discovered when I'm nothing but a stumbling, bumbling puddle of a man, filled to the brim with liquor. It's a reset button, and I need a reboot.

A friend of mine from back in the day is having a birthday party at his house. Good timing. He says come through, says some how do you do's, and put down a few glasses of madness.

A few drinks down, and I'm moving through the party like a hot knife through butter, bouncing from old friend to new before I spot a high school crush of mine across the kitchen, leaning against the fridge.

Maybe it's a misplaced, hormone mixed memory but, to me, she's Aphrodite, a dangerous and mysterious Goddess. A rose too shy to bloom out all her petals.

"Natalia," I say, pointing from across the room. "Who said you were allowed out?"

"What are the odds we'd ever see each other again?" she asks.

We've both made habits of disappearing for years before reappearing without much warning. No one is ever sure they'll be able to catch us or make plans around us because we travel too much, never sit still. We're the same that way. All that matters is that she's here now, with me, and I have quite the buzz.

We chat and drink. She has to leave in an hour to work the graveyard shift at a bakery. After a few more drinks, I throw my arm around her, spilling my beer all over my face.

"Natalia, look what you've done! Zamboni! Lick it clean," I say jokingly.

In a pleasant surprise, she does, and I take it as a good sign that my face is wet from her tongue. Eventually, though, Natalia leaves, and so does my drive to stay at this party. I need louder, bigger.

I start assembling a group and call a taxi to take us to the bar.

I step up to the door, flashing cash, ordering pitchers, Jager bombs, cups of rum. I want it all. I want to feel nothing and be everything.

The night starts moving in a flash, like I'm time-travelling. I'm singing karaoke around midnight. Now I'm shirtless on the dance floor. Good. Never needed a shirt. Never wanted one. Look better without it.

I'm laughing in the bathroom. Another shot. Now I'm at McDonalds. What time is it? I'm sinking my teeth into a fourth McChicken while watching Planet

Earth on my friend's couch. Natalia feels like two weeks ago. I am complete. My purpose was fulfilled.

I drift asleep to the dulcet narration of David Attenborough. "*This particular creature is one of boom and bust ... After months of work, it takes a short break to charge its batteries ... Then the noble pipeliner is ready to go back ... He has a job to do.*"

Christmas and A Long Time Coming

It's been a long time coming for enough things—passion, sex, fighting—so they all decide to come at once. Work finished ten days before Christmas. The boys are ecstatic for two weeks off to spend time with what matters most. For some, it's friends and family. I wish I could say it was for every one of the boys, but some are gearing up for a two-week personal mission against their last two months.

I love this time of year. This Christmas feels special, too. This is not a fairy tale story, but there can still be love at the end of the tunnel.

Me and my Cleopatra have been chatting since we caught up briefly at a party. Back and forth, the slow game of seduction between my Natalia and me is underway. I ask her if she's free tonight for a booze cruise, and she says she is as long as I have her back for work at one a.m. Who knows what would have happened to Cinderella if the prince had another hour to work with?

I miss the streets of Edmonton. The sights aren't

too bad if you don't care about having shit-all to look at. And every man feels more comfortable, breathes a little deeper, when he's back where he grew up. There's a feeling of power and control that comes with something as simple as knowing a shortcut. I am master of my world.

Natalia is a hippy at heart. Wild, knotty hair; skinny, athletic legs; cute ass; face free of makeup. She doesn't believe in it. She's pretty and cute, not beautiful in a movie way, but she has something about her that feels quiet and unattainable. She has the confidence of someone who knows something you don't, like she's in cahoots with God. You want to protect her, but she doesn't need it. Never will.

I knock on her door, and she opens it a crack, apologizing for her dog before anything has happened.

Izzy is a nut case. This dog loves two things: Natalia and balls. She's already freaking out when I sneak through the crack between the door and the frame and, after jumping all over me, she gnaws on her balls to soothe her insanity. If only, Izzy.

My old roommate Carl, is on the couch, limbs spread about the place.

"Bro, it's been a long time," he says friendly.

"Tell me about it, man. Just got back from work. Going for a cruise. How's the back?"

He touches his back reflexively. His back went to shit about a year ago. Not even thirty years old yet. "So much better. About ninety-five percent."

"That's great to hear, man. You've got a few more years in you. You're gonna need your back."

"Unfortunately." He laughs.

Natalia emerges from her room looking like a Salvation Army mannequin—mismatched patterns from disparate decades, holes in her socks. She's covered in Izzy's hair. I wish I could understand why she drives me so crazy.

We open two bottles as soon as we get in the truck. I pass her a bottle. Ladies first.

"So, where to first? Left, right, right, left, right, right, left, right, left?"

"I don't care. Just drive, baby," she says.

The gas pedal hits the floor at her words, and the tunes speak in place of us. We make eyes across the seats and wordlessly make our way through our beers. We drive past the downtown core until the lights of office windows are replaced by dusty, graffiti-covered sheets of wood. We park and watch the drunks stumble through the streets, mumbling universal truths that are forgotten the moment they leave their lips. No matter what you say, where you came from, there's communal beauty here. Each person here has a story filled with truth and pain and suffering, but we look down on them with skeptical eyes. I envy them. Freedom is just another word for nothing left to lose.

"You ever read Charles Bukowski?" I ask. "He's my favourite poet."

A smile fills her face. "I've read all his books!"

"Really? You have them, then?" I can't contain my excitement.

"Most of them."

"Can I borrow them?"

"Only if I have to chase you down to get them back," she says.

"Have you read his poetry?"

She nods her head over the rim of her beer. I pull out my phone. In the dark of the vehicle, the blue light illuminates my face like I'm telling a ghost story. The hum of the engine and faint, near inaudible tune of the radio accompany me like a symphony while I read Bukowski's words.

There's a bluebird in my heart that
wants to get out
but I'm too tough for him,
I say, stay in there I'm not going
to let anybody see
you.
There's a bluebird in my heart that
wants to get out
but I pour whiskey on him and inhale
cigarette smoke
and the whores and the bartenders
and the grocery clerks
never know that
he's
in there.

There's a bluebird in my heart that
wants to get out
but I'm too tough for him,
I say,
stay down, do you want to mess
me up?
You want to screw up the
works?

You want to blow my book sales in
Europe?
There's a bluebird in my heart that
wants to get out
but I'm too clever, I only let him out
at night sometimes
when everybody's asleep.
I say, I know that you're there,
so don't be
sad
then I put him back,
but he's singing a little
in there, I haven't quite let him
die
and we sleep together like
that
with our
secret pact
and it's nice enough to
make a man
weep, but I don't
'weep,
do you?

We stop and talk in front of her building when I take her home at the end of the night. She asks why I chose to read the poem I did. I shrug and say I like it. She stares into my face, and I feel seen. She says she likes it, too. She wants to hear it again.

My Brother

I swear, my brother and I have been fighting ever since I was in the womb. He's my older brother. We followed similar paths—he's a power engineer in Fort Mac.

We've been at war for as long as I can remember. It's normal for brothers to fight, but we've been in a power struggle. I got bigger, stronger, smarter, better in things that a little brother shouldn't be better in than an older brother, and it was something he couldn't handle.

When he was at his lowest, the boys and I had his back, bought him beers, offered shoulders for his tears, words of encouragement to keep his feet moving. We stood by him and only wanted the best for him.

When he got his job as a power engineer, which was a far better job than ours, he made sure to tell us that we were below him, that he made more money than us, that we were nothing but a joke, dirty rig pigs not worth the air he breathes. I couldn't believe he would spit in his brother's eye like that.

"You are not a brother of mine," I told him and walked away.

That was the last time I saw him, nearly two years ago.

It was messing up the balance of our family. Fortunately, or unfortunately—however you view it—the universe craves balance.

I am at the birthday of my oldest friend, Jeracles. He got together all the good, normal folks I grew up with. I call them Disneyland people. Their life is a happily ever after story.

It's actually nice to see a bunch of fine people when I get home. They talk about family and work and the weather, what television to watch, what movies to avoid. It's boring as fuck, but it's a nice change. They are respectful and are careful to be polite, to never offend anyone, as rocking the boat is worse than never leaving shore. To be honest, I love these people, but I'm not one of them,

"Eh, Jer Bear! Happy birthday, buddy!" I yell as I walk into the house.

"You made it! I thought you were working." He runs and jumps into my arms, and I spin him around in the doorway.

I weave my way through the house, stopping every few feet to hug someone, shake hands, make quick, friendly jokes. I see Natalia, ripped jeans and a sweater, standing in a group of barefoot women in the living room. I play it cool and catch up with others in the kitchen party.

A few beers down and someone I somewhat know surprises me. He leans in close, clearly uncomfortable with what he is about to say.

"I heard you can get things …" he says in barely a whisper.

I chuckle. "What is this, *Shawshank*? What you need?"

"I was wondering if I could get something for me and my wife. We've never tried MDMA." He looks over at the group of women, and it's clear the woman staring through the crowd directly at us is his wife.

I smile at her.

"Who the fuck do you think I am? Some kind of criminal?"

He takes a step back, eyes wide. He stammers out an apology, and I break out laughing.

"I'm messing with you. Here, come to the garage," I say. Still laughing, I pick out a rock of MDMA from my pocket and hand it to him.

"You scared me," he says. "How much do I owe you?"

"Nothing, man. You owe it to you and your wife to have a good time." I slap him on the shoulder. I like helping people.

When he exits the garage, I pop some into my own mouth. I like helping myself, too.

Back inside the party, I head right towards Natalia.

"You work tonight?"

"One a.m., as always," she says.

"Fuck that. You want a little M?"

She looks at me with a devilish smile.

I place some in her hand, and she washes it down with my beer.

I get a ride with temptation when it's time to head to the bar. She drives a beat-up old Ford Ranger, two-seater, both ripped and duct taped. I remember we parked and just sat in the truck, luxuriating in the tension between us, inching closer to each other. My mind was going blank, and all I wanted to do was kiss her long and softly. Then, suddenly, we're in the sharp December chill, walking to the bar, and the top of my head is buzzing, pulsing pleasure down to my feet.

I did nothing! Like a fucking coward! And I barely care!

As the beers go down, I start feeding people rocks of MDMA, feeding myself a little each time. *One for you, one for me. One for you, one for me.*

The rock I gave that guy and his wife was split with another friend before I even started feeding them more. So, half of us are sitting in the bar, talking fast and sweating loudly, higher than the heavens. Our conversation is mostly about how amazing we feel, and it still feels like the best conversation I've ever had.

I slip and skip through time for a bit until I'm alone, walking down Whyte Avenue, melting into and with the music from the buskers and the energy of the people. Someone yells at me from a line waiting to enter a club. It's my brother, Marko. It's been two years since I've seen him, and he catches me when I'd befriend a blank wall. *Thanks, Universe.*

"Hey! How have you been?" I say with a smile and a huge hug.

"Good. What are you doing?" he asks.

"Ah, you know … getting a little too rowdy," I say, patting the sweat from my forehead.

"Fuck, really?" he says, laughing. "Well, fuck,

let's go grab a beer. Let's talk, man!"

We walk down Whyte, skipping bar after bar. None feels right to either of us for whatever reason. Too many stories between the two of us to enter a bar without the proper energy.

He tells me how he got into stocks and cryptocurrency. I tell him about Ecuador. The two year gap between us isn't mentioned, but it hangs on the edges of our conversation and in the hesitation before questions like, "are you still with so and so?"

He invites me back to his place for some whiskey.

We arrive at his house, and I put more liquor into my already poisoned body.

Whiskey is a bad choice when resentment lives in the room. Anger starts to stir. I don't know who made the statement that first started the flow of emotions, but it doesn't matter. How do you blame a single raindrop for the flood? The levees are down, and the past is spewing out all over the table.

"You told my best friend about the penthouse, man!" I say in my drunken voice. "I asked you to tell no one."

"What the fuck is the big deal? I'm proud of you, fucking proud," he snaps back.

"That's not the point. Not the point. I asked you not to tell a single soul."

"Fuck you," he says, dragging out the curse. "I was proud of you, so I told your fucking best friend. Who cares?"

"Don't you think I should have been the one to tell him? The guy was choked that he heard it from you and not me. Don't you understand? Don't you understand people aren't happy when you have things?"

"If people are like that, then fuck them," he barks.

"What do you know about how to treat people?" I say.

We pour more whiskey to continue the negotiations. We both know all this is coming out on the table, all our issues, all our past. It wasn't done. It hasn't even started. Tonight, we'll sort out our problems or die trying.

"What the fuck, you motherfucking dog?" I shout whiskey curses. "You dirt. You dog. You spit in my face, the faces of the boys, the ones who had your back, who heard your tears; the only ones who gave a damn about you. Then you fucking tell us that we're less and you're more?"

"Yeah, you're right. I fucked up. And I'm sorry," he says. The tears in his eyes match mine. "I wish I didn't say that. I wasn't in a good place. I wasn't a happy person, and I was in a low place.

"All you guys were killing it. Making good money while I was in school, poor, nothing, and fuck, the job went to my head."

"We never said a word about you like that," I say.

"You didn't have to say anything! It was obvious. I heard it every time you said you'd cover the bill or Wilton told me not to worry about it."

"So, instead of buying the boys a few beers, sucking it up and saying thanks, you go and chop them down? We're all dirtbags, you included. We're all trying here," I say. My eyes glisten, fists clenched below the table.

He drains his glass and pours it back with whiskey. His voice is confident, definitive. "I said I was sorry, and that's all you're going to get. That's it. I fucked up. You can take it, or you can fuck off."

"That's it! Get the fuck outside!"

Words don't solve everything. Only fools and cowards think they can. It's time to fight. Go bone on bone. Connect fist to flesh. Battle cries and blood.

Outside on the cold concrete and thin layer of ice

covering his driveway, we circle each other. It's somewhere around four in the morning.

"How do you want to do this?" I ask, looking for the rules of engagement.

A left hook smashes directly into my nose, making an audible *crunch*, before I truly realize we're in a fight-fight.

Blood trickles into my moustache, slides down into my mouth. I taste wet tin. I love it. I feel alive. I want to destroy something.

I run at him, duck a hard left, and get behind him. I drop hammer fists into his skull. He tries to get away, but I keep my feet quick and moving, and I don't let him out of my reach.

He's off balance on the icy cement, and I use that to swing him around and throw him down the length of the driveway. He slides on his stomach like a penguin until he's near the road. I dive on him and grind his skull into the sidewalk with my forearm.

Somehow, a rogue fist of his finds its way into my eye socket and rolls me back until he's able to kick out from underneath me and stand again.

There's a shooting pain behind my eye, like a strobe light. Give me another one. I love pain. I live for it.

He charges me, and I hit him with a knee and flip him over my hip, raining fists down on the back of his head until he yells *stop*. I respect the terms of his surrender.

"Fuck, you're fast." He chuckles through a cough, flat on his back on the driveway, feet sprawled out on the sidewalk.

"You smashed my nose good," I say.

"I love you, bro."

"I fucking love you, bro."

We hug and embrace.

"Goddamn. Who hammer-fists? God, that hurt. I thought I could take it, but they just didn't stop."

"Yeah, you got another pop in my eyeball. That'll be a good shiner. Chicks will dig it." I laugh.

"So, we're done!"

"What's done?"

"We're done." His face is angry again.

"What's done?" I yell back, angry for some reason.

"We're done, or we're doing it again, fucker."

There's pure rage again in my eyes. I taste bile on my tongue. He feels the same. We're two raging bulls waiting for the gate to open to gore the other.

I was born for this; born to wield a sword and axe

on a battlefield, swinging it into another man's chest and using my leg to free it for another swing. Templars. Warriors. Gladiators. I don't belong in this present-day society of limp dick rules. These fragile, basement dwelling sheep. I've always wanted truth, honour, and there's no truer way to learn about yourself than combat. It's fucking beautiful.

We square up again. This time, my Muay Thai kickboxing and karate training comes back. I fake a jab before delivering a shattering side kick he doesn't expect. He stumbles, rattled. I repeat, connecting again. Then I dodge a hook and bring him to the ground and treat him to another course of hammer fists. I feel my hand break and feel nothing but pleasure. I am nothing but an animal. I am primal.

"Okay, okay, stop, stop," he says.

In a rage, I give him one last kick into his ribs. He yelps in pain.

"We're done," I say, looming over him.

His breaths are ragged and laboured. He looks up at me, clenching his side. My fists are still cocked. "We're done," he says.

I lift him from the snow-covered ground. Then I hold the door open for him, and he limps inside, holding his ribs.

My hand leaves a bloody print on the door, and I smile as I watch the wet red slide down the glass pane. I grab a glass of water for him on the couch, tossing him an ice pack and some painkillers.

"Thanks," he groans as I make my way to the bathroom.

In the mirror, I smile at my blood-covered face and teeth stained maroon. I push hard on the bruising forming around my eye, already a shallow purple. I'm beautiful. My right hand is swollen, dark blue, and my pinky stands crooked. Perfect.

The hot water of the shower lifts the blood from my skin. The water down the drain looks more like diluted blood than H2O. I don't know if it's mine or his.

I'm under the water for a long time, imagining I'm surrounded by liquid but can still breathe. I picture myself not drowning until I'm clean and the water runs clear off my bruised body.

When I leave the bathroom, my brother is still on the couch, ice on his head, eyes closed, mouth open to take in oxygen.

"How are you?" I ask.

"Fine."

"Need anything?"

"Ice."

"We good?"

"We're good," he says.

Twenty-nine years, that storm was brewing, and tonight, it touched down. All our youthful hatred, adolescent fascination with pushing each other's buttons, the power struggles of our twenties, stirred and shifted in an earthquake, a hurricane of two brothers in the night. We left it on the field. We bled out all our problems and wiped the slate clean. Blood and broken bone and cracked ribs was our detox. We are now pure.

The next day will kill, the next few weeks will hurt, but every sting of pain will remind us of our truce. The bones will heal and strengthen our bond, and we will always remember the day we finally became brothers.

Natalia

The black eye is a sickly yellow come New Years. I'd rather it had more staying power. Now I look less like a warrior and more like a kid with jaundice.

Wilton is all flanged up for the New Year's party. He's hoping for a wild one to forget the last year and welcome the future, all drunk and wild and full of insanity covered in chocolate sweets, sweat, and sex.

"When tomorrow gets here, where will yesterday be?" he asks.

"It'll be gone," I answer.

"That's right," he says, smiling. "It'll be fucking gone."

I message Natalia to join us on our night of adventure, drugs, and madness. She keeps me in suspense until the eleventh hour, then responds only with, "*ok.*" Good enough for me. The night is set!

"Buddy, I can't wait. We going to Petey's!" Wilton says.

I'm fired up. I love it. I can't wait. So pumped.

Beers with my best bud. “We’re going to get lit.” I say while jumping and softly check him with my shoulder.

“Is the girl you’re chatting up coming?”

“She said she is. But who knows? She likes to flake.”

“Whatever. We’ll have a blast.” Wilton rides an imaginary horse around the house.

We arrive at Petey’s house party in St. Albert, a town named after saints corrupted from the inside out. Two floors, one kitchen, pool table in the basement, beers in the fridge. We head downstairs.

“Hey, don’t let Alice come in this room.” Wilton looks at me seriously. “Okay?”

Alice is his girlfriend. She’s got her head screwed on right; therefore, she’s not a huge fan of him chopping lines. But see no evil, well, then there is no evil.

“I’m on it,” I say, shutting the door behind him with a wink.

As if on cue, Alice descends the stairs, looking for Wilton, spots me, and dances over.

“Alice, Alice, Alice, what you been up to?”

“Just drinking,” she replies, giggling.

“You’re not drinking,” I protest. “Let’s go

upstairs, and I'll show you drinking! I'll let you choose whichever shot we start with."

"Tequila!" she screams.

"Fuck. I hate fucking tequila."

"I knew you'd bitch out. I could hear you bitching out from the top of the stairs."

I look at the door behind me. It's closed, and it's going to stay that way. "If that's what you want, then let's do it."

To the kitchen we go, and I put that toxic Mexican poison in my system, all so Wilton can do a sneaky little rip with the boys.

As I'm washing my mouth out under the tap, I see my Goddess has arrived in the kitchen. She's wearing a baggy T-shirt, half-tucked in tight, and faded black jeans.

"You made it!" I say, sounding a little too excited.

"Better late than never," she replies. "So, I brought you some beer to apologize."

We each crack one open.

Before I even take a sip, she leans in and whispers in my ear, "you want to do some drugs?"

I think I'm in love.

She pulls a clear crystal bag from her pocket.

"Look, I got the good shit."

"Thank God. I've only got this dark shit that gives you a hangover," I say.

"What would you do without me?" she asks.

Lately, I don't know.

We split hers in half and wince as it hits our tongue and burns our lips. It's disgusting.

We kill the taste with beer and wait.

Slowly, the happiness starts to fill my brain and extends through my bones and tissue, forming an aura of pleasure around my body.

The beers are doing their job. The drugs are doing their job. The great company is doing their job. I'm in and on ecstasy, and this New Years is perfect.

We congregate in the living room and watch the timer on the television until it strikes midnight. There are cheers and hoorahs, and instantly the sound of fireworks pervades outside, followed shortly by the barking of neighbourhood dogs. People hug each other. People kiss each other.

"It's bad luck to not have someone to kiss," I tell Natalia.

Fireworks, cheers, electricity when our lips meet. We kiss, and the moment stops. I don't hear anything around me, yet I'm surrounded by noise. Our lips touch; wonderful, wet lips become tongues; hands grab hair and pull towards each other's bodies. Eventually, we pull apart.

It's fifteen minutes past midnight. The party has continued on, out of the room.

We stare at each other, somewhere between awkward and aroused.

"Well, that was lovely," I tell her.

"Ha, yeah. It was nice."

"You wanna get a shot?"

"I would."

As the night goes on, I lose motivation in the party and the loud music. I want the room empty except for Natalia and me, the air feels thick with the promise of pleasure. It's like she can feel me thinking about her, because she sneaks up behind me and whispers in my ear. Her warm breath and soft voice almost finish me off without any need for touching her.

"I have no one at my house," she says.

"I'll get a cab!" I nearly elbow her in my hurry to grab my phone.

We arrive at her house, and her fucking dog is jumping, howling, biting a tennis ball, launching itself around Natalia's apartment, barking, whining. Izzy lives for her and, in any other context, it would be kind of cute.

We go to the couch and, with a few button-clicks, soft music begins playing.

"You know what I want to do right now?" I say over the soft music, trying to ignore the German Shepherd in my line of sight.

"What?"

"I want to take the creamiest body butter you have and cover every inch of you with it. Do you have some?"

"Yes."

"Can you go get it?"

"Yes."

She disappears somewhere in her apartment.

Anticipation. I close my eyes and roll back into my mind, the drugs sparking a pleasurable confusion.

She returns with the lotion and sits back down beside me.

I slowly take off her top—no bra. I peel off her jeans to reveal pink and black lace underwear. I leave those on, still needing something for the imagination.

"Lay down here."

She does what I ask, not taking her eyes off me as she does so.

I take the cream and drip it slowly on her body. Her body writhes in pleasure to the cold temperature, her eyes close, and she sucks air through pursed lips.

I run my fingers along her spine before running my hand lower to massage her long, slender legs. Gently, I touch the inside of her legs, and she spreads them for me, so I move higher, all the way up to her pussy.

I softly roll my thumbs into her wonderful ass cheeks and linger there. The music plays. Her underwear is getting in the way, so I tug at them to signal they need to go. I pull them down to her knee, she kicks her feet up, and they slide off. I throw them over my shoulder.

More cold cream drips on her body. I caress her ass cheeks, buttering that ass up good. I can tell she's excited and moist and unsure what will happen next.

I glide my hands along the curves of her body, her neck, her collarbone, all the way down to her hips. I'm in no rush. I make sure not to touch her where she craves, not yet.

I put my body over hers, my fingers through her hair. She smells intoxicating.

The goddamn dog does not like what's happening. Izzy is jerking her head around, biting that fucking tennis ball, howling. She doesn't want to share Natalia. And neither do I!

I flip her over and put my hand around her neck, the hint of danger, and then I slide it gently towards her cheek and kiss her again. I move my way down and kiss her neck, then slide my tongue in her ears.

More cream.

I lick and suck her tiny tits, circling my tongue around her nipples. Finally, I make my way south and give a long, deep lick of her pussy. I can promise you that she has one you want to lick until you get to the centre.

I take my finger, enter her slowly, and massage with the come-hither motion.

The fucking dog has now positioned itself on my legs and is really making things difficult. It's hard to move into a more comfortable position so I can penetrate better with my tongue. Well played, dog (you bitch!). Always jerking her head and growling, she takes the ball, presses her nose in my ass, and chews the ball against my hole.

"Ah … stop that! Get out of there, Izzy! You don't know where that's been," I yell at the dog.

"Izzy, fuck off," Natalia barks.

I pull down my pants, making this couch

complete—two naked bodies and one insane dog. I enter her. It's like drilling in a swamp. I mean, she's wet.

Our two bodies entangle, hot, sticky, and sweaty. The thrusting starts slow and intimate. I'm going to take my time and enjoy this Goddess. Steadily, the slap, slap, slap gets faster and faster, and we start to grab and pull closer and deeper.

Then the dog jumps on the couch again and firmly plants herself on my legs.

My thrusting power has been reduced. I'm at fifty percent power! Fine then. I like a challenge. I'll thrust harder. I'll make fucking sparks. Goddamn, I'll rock her like a goddamn fracking drill. You'll see, dog.

Again, the nose and ball biting presses into my ass.

Get out of there, I think to myself. In the end, though, I just accept it. It's weird, but not the weirdest thing that's happened to me. But fuck it … pound harder, thrust deeper.

"Get on top and ride me like a cowboy," I say.

The dog's behind the couch, jerking its head, growling. She doesn't like what is happening.

Natalia bounces up and down. We're loving how

drugs make us last forever, and it feels great.

Eventually, we are both propped up behind the couch in standing dirty dog position. This is fitting, because there's two dirty dogs here. Me and wherever Izzy's nose has been, other than sniffing my ass. You can hear the slap of skin on skin, the moans of pleasure. I have one hand full of her hair and the other around her belly. More moans.

This fucking *dog*! It's distracting. It jumps and headbutts my thigh and bare ass, growling over the music, clawing at my legs, thrusting.

I think I'm going to finish. Harder. Deeper. Faster. It's time.

I pull out and, in my excitement, all my white, sticky cream covers everything. Her bare back, the couch, the floor, the ceiling.

Fucking dog saliva all over my legs and ass, scratch marks everywhere from her claws.

We both roll onto the couch, exhausted. She's in my arms. Our minds are calm and clear. Nothing else exists, nothing else matters beyond this second, beyond the next. We listen to the music until we eventually drift off into the nothingness of sleep.

Booby's Back

We went back on the fourth day of the new year, part of a broken crew again. Wilton got a better job inspecting and, like the little slut he is, went to a different pimp. Chiggins got sent to another crew, too. I'll miss the little wart. Now I was temporarily without a crew, without a rig family.

I know oil and gas. We get shipped to one camp then to a different one, waiting until we're oiled and gassed again.

A rumour started to spread, and it was growing through the crew line that Booby Mac was returning from the bender to end all benders.

When I finally see him, he looks like he's been dragged out of a river. His face hangs down, skin stiff and worn like an old catcher's mitt.

"Booby, ya dog. How are things, my friend?" I say.

"Terrible. Just terrible," Booby says. "Getting a fucking divorce. The bitch is such … She's such a … just a fucking bitch. I don't know."

“Maybe it’s for the best. Mexico didn’t fix it?”

“Oh, Mexico.” He shakes his head. “Terrible, just terrible. The family didn’t know we were fighting, so we had to fake being happy and cuddly the whole damn time. The bitch got so drunk, so I got drunker. No sex. Just terrible.”

“No sex? Not one shag on a Mexican vacation? That’s gotta be a first.”

“Well,” he says, rubbing his palm along his jawline, “I guess I’m a pioneer.”

“So, what’s going on then?”

“One word. Divorce,” Booby says. He stretches his arms high above his head, and his belly protrudes below his shirt. “We get back from Mexico, and she kept making my life so miserable. One night, I just said, ‘Fuck it, we’re getting a divorce.’ You know what she said back to me?”

“What’d she say?”

“She didn’t fight. Didn’t ask any questions. Just said, ‘Yup.’ Isn’t that cold? That’s so damn cold.” He’s pissed again, lost in memory. “Like she was making my life so terrible so I would demand the divorce. I’d still be the bad guy. They’re smart. Smart and evil women are all I seem to find.”

“What have you been up to besides that?” I ask.

"Well, after that, I started drinking. A lot. I mean, heavy. Like, I go to the Colehearst Pub and drink a forty-pounder by three p.m. Then I drive home, drink some more, pass out in the hot tub, surrounded by cans. Then I wake up, do it again. Don't even think I really get that drunk. I just get … numb."

"What do you do at the pub? Just drink alone?"

"Yeah, I just get there, sit down, and tell them to leave the bottle. Drink it, think, pay, drive home. Every day."

"Gotta say, you're depressing me here."

"Ah, shit, sorry. I bought a brand-new snowmobile. That's something fun, eh?" he says. "That made me feel good for a bit. Went to Fernie with the boys, got my mind off the divorce for a while."

"Didn't you just get one last year?"

"I sure did," he says with a chuckle. "This new one cost me sixteen-thousand. All the bells and whistles."

"Sixteen-thousand dollars on a fucking snowmobile? Are you nuts?"

"Anything so she doesn't get any more money out of me," he says through clenched teeth. "I've burnt through about forty-five-thousand dollars in the last month and a half."

"What?"

"It hasn't been good. My other buddy Rubinho is getting a divorce from his wife. And his wife is my ex-wife's best friend. They're a real treat of a tandem."

"You're a living soap opera," I say.

"Preach. So, me and my buddy go to Vegas for New Years. We got off the plane and went right to the strippers. I get in there and demand blow. Demand it! And I demand two private strippers. And the blow is shit, so I say get me the good shit or stop wasting my time."

Christ, Booby. It's like he lived the last two months like someone who got to be him for a day. No regard for tomorrow. It's fucking beautiful.

"So, for two and a half hours, I had these two strippers mashing their tits in my face, while I'm doing lines off titties, lines off their asses—all of it. Finally, Rubinho comes and gets me. I'm fucked. Just fucked. Can barely see. I couldn't even read the bill. I just paid whatever they said. I wake up the next day and look at my account and find out that it was twenty-five-hundred dollars in lap dances!"

"You're fucking insane. Insane," I say in

disbelief.

"Yeah, I know," he says. "So, get this. I see my divorce lawyer two days later. She's asking me about my finances and if there's anything I should tell her that's weird. And I tell her that, on my bank statement, there should be a bill from a strip joint called Little Red Devil. You should have seen her face, bud. Priceless."

"Well, not priceless. It cost about twenty-five-hundred dollars," I point out.

"She asked me: what does twenty-five-hundred dollars get you at a strip joint? And I look her dead in the face and reply … Memories."

The Pickle

There are people who believe in good and evil. I don't agree with that, about morality as different sides of a line. What's stopping someone from stepping over the line? Lines are only made to cross, then you draw a new one. It's all perspective. You make your own good and create your own evil. It's a personal invention that shifts day by day, changes with the season and the tides. I don't believe in good and evil. I believe there are strong people and weak people.

Humans are fucking with nature by letting the weak survive. There are too many goddamn people. We live too long, too easily. A sick grizzly bear doesn't mate; it dies alone in the forest to be eaten. This isn't a sad scene. The opposite is a sad scene, where the failures can move through life, still getting something for nothing. This is a plague.

Luckily, out here in the bush, there is no good or evil. Only the strong and the weak.

I'm on nights with a bunch of solids—Booby Mac and Micky. We got Special Ed, too, but I've come to like him. He's a simple man, but he's still here, and when I ask him to do something, he does it (within reason).

I wrote a poem about him.

How?
How did this human gene-type survive
the test of time to produce a Special Ed ...?
How did he survive Sabretooth Tigers?
How?
How did he survive cannibal tribes
that shrink your head
to the size of an apple?
How did he survive tiger tanks?
How has he survived to today?
For the sake of humanity
I slowly let him
trust me.
Then, I'll strike.
It'll be quick.
I'll be the unsung hero of humanity

Then there's the Pickle. He's new. Half his face is hidden behind a dense red moustache that he's been growing since the mid-90s. He smokes two packs a day, and the teeth remaining in his mouth are stained yellow, decaying little chicklets, like a pumpkin in October. He wears his years in the deep, canyon-like grooves on his face, and his head has a shock of red hair.

I personally have nothing against gingers. It's kind of weird I felt the need to mention that. It should be normal that I don't have anything against them. His skin is luminescent, and his body so fragile that I'm sure he would break if a child shook his hand.

I introduce myself with a wave.

He looks at me through his eyebrows and decides to engage.

"Nice to meet you. I'm Dan," he says.

"How long you been doing this? Your face looks lined with experience."

"Since I was sixteen," he says.

"Where you work before this?"

"I was an S and S man."

"Ah, you worked for ol' Snort and Sniff!"

S and S was a company of cocaine cowboys. Real wild boys who abandoned the idea of sleep. They would work for days to earn their party.

"There were some wild years," he says with a laugh.

Like always, the first few days, things move normally. Guys get settled, acclimated. It only took a week for Dan to show his true colours. Fucking gingers. I don't have anything against them, though.

He would go for water, which should take a cool hour, but he wouldn't roll back for nearly three. We were waiting, and he was slowing down production. We checked the local watering hole and followed his drunken, staggered snow path like bobcats hunting innocent tourists until we came across a little pack of empty PBR cans lightly covered by fresh snow. Pickle was out drinking, wasting our time. And not even inviting us!

I open his crew truck door and right on the passenger floor is a fifteen pack of PBR's, ripped open at the handle, and a bottle of dry London gin. So, we crack on.

I don't say a word, but there's an inner anger building.

We're doing a long shot under a river, which means we need technology. Specifically, a hyper precise GPS system that uses satellites and the earth's gravitational pull to get data. We need to cart these heavy-ass pieces of steel in the shape of a drill pipe. The stem is attached to a tri-cone drill bit. The tri-cone drill bit in the front does the drilling, which is then threaded onto a piece of drill stem with a two degree bend to it—this is how we steer—and that tool is then threaded onto something we call a gyro. Then our own drill stem is attached, and now we are drilling, baby. This darling is accurate to the centimetre.

Of course, the technology needs a power source. So, let's say we are drilling an eight-hundred metre shot. On every rod connection, we'll need to splice wires. If a rod is six metres long, we add six metres of wire inside the drill stem. Then the next rod goes on, another six metres of wire is spliced, and now there is twelve metres of wire, twelve metres of drill stem.

At the end of the shot, there will be eight hundred metres of wire inside the drill stem, giving this piece of technology its power. It's called wire lining. The splices are connected with little metal tubes crimped together, and then we put on shrink tubing over that. The tubing is to repel water from sneaking inside and shorting our connection. That's how we supply power to the gyro so we can steer underground.

The inspector comes cruising around while checking the site then comes out to chat with the boys, as you do when you're an inspector.

“Hey, boys. So, how are things going?” the inspector asks.

“All good. Just trying to get this pump going,” I say. “Eddy, go grab me a crescent wrench, bud.”

“Which one’s a crescent wrench?” Ed asks.

“What? The one you can adjust with your thumb,” I reply.

The Pickle comes towards us to grab some new gloves out of the crew truck. He opens the door and, right smack-dab in front of the inspector, reveals an open fifteen pack of PBR's.

I quickly look at the inspector with bewilderment. Luckily, he doesn’t see them, or doesn’t want to. I ninja-kick the door shut.

“Hey, what the fuck! I’m grabbing some gloves,” Pickle yells.

I motion my eyes towards the inspector.

“Ah, yes, must be on the other side.”

When the inspector leaves, I storm around the vehicle towards the Pickle.

“What in the fuck are you doing? You open the door in front of the inspector and there’s a whole case of beer and a half-drank bottle of gin!”

“Must have forgotten about that,” he says nonchalantly.

"You've been doing this a long time, man; I get that. S and S are some wild boys. But that kind of shit will get the whole company kicked off site, black-balled instantly."

"Chill out," he says through his moustache.

"You selfish prick. There are other guys happy to work here, and you are going to fuck the whole crew because you can't wait a few hours."

"Chill out," he says again.

"Hide that shit," I say. "You understand? Get rid of it."

Later, the inspector comes up and motions me over for a little chat. He takes me under his arm and speaks in a low voice.

"I get it, bud. Some people are drunks. But you can't have that shit in front of me. If I see that again, you're all gone. No second chances."

"You'll never see it again," I reply through clenched teeth. That's a mark against us, a stain from Pickle's filth.

I talk to Booby Mac. He tells me to get rid of him.

The Pickle has Crohn's disease, and it's acting up. Apparently, it causes a person to shit blood. I wish I had never learned that. But he doesn't control it much, doesn't eat healthy or treat his insides well. All he does now is get shitfaced at work, take bloody shits, and pop morphine pills like Skittles. Every meal becomes a strict liquid diet, and he's wasting away, tissue and bone and thin muscle bound in a bag of dusty flesh.

Pickle is in the truck, feet on the dash, complaining of stomach issues. Again, the weak getting cared for by the strong.

My palm slapping against the truck window startles Pickle from his drunken slumber. I motion for him to roll the window down.

"You useless little man," I sneer. "Just sitting in the truck while we work, eh?"

He sparks a dart. "What'd you say?"

"Get the dicks out of your ears, you blood-shitting fuck. Why are you out here?"

He exits the vehicle and tries to make himself big. All of his one-hundred-and-twenty-six pounds of unintimidating, wrinkled ginger-ness squares up in front of me. Maybe there is a little fight in him, after all.

"What the fuck you say to me?"

"Listen here, bud; you've been sitting in that truck, sipping warm beer, smoking darts, for the last two days, not doing a goddamn thing. Open liquor in front of the inspector? Day trips out to the water hole? You selfish little fuck. Why are you even here?"

"Oh, fuck you, skippy. I've been shitting blood for the last three days, waiting for someone to change shifts with me."

"That sounds terrible. I don't envy that. But you're not helping us out here. If you're that sick, then go home and see a goddamn doctor. You ain't helping us," I say.

Booby can see the scene from the drill—a couple of boys hashing out a problem with finger pointing and hands flying above our heads.

"I'm sick! I'm shitting blood. I'm going to be out of here soon. Don't bunch your panties."

"All you've done is smash beers and pop pills inside this goddamn truck! It smells like a fucking brewery in here. The Crohn's sounds terrible, but you can't be handling it by doing this. It's not doing you any good, and it's not doing us any good. Call Rossko," I say sincerely.

"All right."

"All right," I say.

We leave better than we started. Not friends, but on common ground.

I walk up to the drill.

"How did that go?" Booby asks, laughing.

"Nothing but a blood-shitting, pill-popping skeleton of a man. Weak. I told him to go home."

The Pickle left the next day and was never heard from again. I like to think I was the reason.

The weak and the strong are how life is divided. A newly crowned lion kills all the cubs that are not of his own blood to assert his rule, cement his dominance. The lion is strong, and the cubs are weak. It's that simple. Neither is good or evil. It's all perspective.

One-hundred years ago, Pickle would have died penniless in the streets. He would not be able to ride on the shoulders of the strong. I wish I could just take him out in the jungle, far away from civilization and creature comforts, and let nature go to work. The oil field has no patience for weakness.

Minus 52

Canada has a lot of oil, and Alberta is home to the third largest oil reserve in the world. It's said that Saskatchewan might have more oil than 'Berta, which would make Canada the biggest oil capital in this whole goddamn world. Venezuela has the top spot, followed by Saudi Arabia. Both of those countries are hot and sandy, and people don't even need to wear clothes if they don't want to. Wouldn't that be nice? Lucky bastards. I wouldn't mind that, especially as I blow hot air into my wool sweater and hold my hands near my balls because they radiate heat.

"Micky, my man. We got a shitstorm coming our way, eh," I say, wide-eyed, teeth chattering.

Micky casts a nervous eye skyward. "Buddy, it's going to be rough. Fifty below. Don't let your skin show."

Luckily for us, we're on nights, so we're set to take the full force of what nature can throw at us.

The cold snap is set to hit our crew in about twelve hours. Day shift workers take their posts and prepare for the inevitable battle that will soon be upon us. There are a few things that can't fuck up, that we need to focus on during this wave of freezing.

The drill might not be able to work in this temperature due to flash freezing. As soon as the drilling mud hits the drill stem, it flash-freezes and turns into solid ice. This is the worst-case scenario, but it isn't the only possible negative scenario.

The mud tanks must be constantly monitored, and the Hotsee—the pressure washer—is our necessary ace in the hole. The Hotsee is the washer that boils our water so we can heat frozen shit, clean the screen, and clean the mud. All of these are necessary to keep us working.

"Listen up, you two," I tell the Cole's. "A storm is coming our way. You have to make sure the Hotsee and tanks don't freeze, all right?"

"Yeah, yeah, yeah," Cole answers in his low voice. "I'll make sure she's good when you come back."

I have dreams of stupidity. Idiot dreams.

The cold snap is hitting hard and fast, and it hits early. When we went to bed, it was a perfectly bearable negative twelve. When we woke up at five p.m., it's already thirty below and falling by the hour.

We hope the day shift had their shit together so we can actually get work done.

"Fuck," Micky says. "It's so cold that I didn't even go for a smoke when I woke up."

"This weather is cruel."

"Oh boy, it's a beautiful day. Just a beautiful day!" Booby says, rubbing his hands together in anticipation.

We throw the nearest object we can reach at him.

"You're weak, boys. Weak, I tell ya. This isn't Panama or wherever you own land. This"—he breathes deeply and smells the air—"this is Canada. It's a great day for drilling."

We arrive on site to find it in full shutdown. Too cold to drill. So, we're on rig watch, tasked with making sure shit doesn't freeze on orders from Calgary.

I walk towards the tanks with the wind painfully burning any exposed skin. Near the tank, I can hear the Cole's talking to someone.

"What's going on with the Cole's?" I ask one of the guys nearby.

"He's fucked, man. Froze the Hotsee and been up there running around, talking to himself for hours."

"What the fuck do you mean the Hotsee is

frozen? What the hell happened?"

"Guy didn't check the filter."

"Get fucked," I say.

I sprint up to the Hotsee and see the skinny pressure hose frozen stiff. All I hear in my tank is the rambling of a madman talking to himself.

I climb the ladder to my tanks. "Cole, what the fuck, man?" I look down at this poor, confused man running to and fro with five feet of hose in his hand like it was his dead pet. He's weeping, cursing, and ranting.

"I'm going to quit. I'm done. I had no one helping me. Everything has come undone, and no one helped me," he says in his low voice.

"Just tell me what happened."

"The Hotsee just quit. I don't know what happened," he says in his high-pitched voice.

"Did the filter get plugged?"

"Yeah," he says.

"Cole, you just fucked me," I say softly. I know it's a delicate situation, and he's frantic. "Man, it's negative forty right now, and we have no Hotsee."

"I'm such an idiot. I'm such an idiot. I'm going to quit. I'm done. I'm done. I can't," he says.

"Chill out. I'll take care of this. Don't worry. You

fucked up. But is everything else all right? No bullshit. I need to know everything that's wrong. I can't have surprises."

"I'm just so stupid," the Cole's yells. "Fuck!"

I take the hose out of Cole's hands and give him a hug. He's still shaking, halfway into a mental breakdown.

After the hug, I tell him to get in the truck and get some sleep, that I'll take care of it.

The day shift leaves, and the night shift's battle against winter begins.

The wind is screaming in my ear, and flurries are whipping through my wool jacket. It feels like my blood has been replaced by ice, and my heart is pumping crystals through my veins.

Micky emerges through the haze of white with snow hanging from his eyelids. Snot runs from his nose. It's stained the back of his glove from successive wipes.

"Micky! I need your help with the tarps," I shout above the wind. "Cole's froze the damn Hotsee. The hoses are cock stiff."

"Ah, you're kidding me."

"Wish I was. It's a shitshow. Can you help me? We got to tarp the thing and get heat on this thing. Get the Herman Nelson."

He grabs the tarp and drapes it over the Hotsee. He's pissed. I'm pissed. Yet, that changes nothing, so he secures the tarp with bundles of snow and goes to get the propane heater while I go to find propane tanks.

Damnit, the Cole's fucked us good.

This cold is unbelievable. My breath lingers like little crystalline clouds in front of my face. I could breathe my last breath, and it would hang in the air for eternity while I crumpled to the ground to be quickly buried by falling, blowing snow. The trees around us use the snow to make blinding white armour that clings to them like a dying whisper. All I hear in the distance is silence, sometimes pierced by the scream of diesel motors begging to die, because the oil has turned thick inside its arteries. I know I'm human because I can keep going, even when I shouldn't.

"Micky, you gotta go warm up. There are icicles hanging from your eyebrows. I'll take my shift out here," I say.

We need micro-breaks to warm ourselves. It's a full-time job just to make sure everything stays running; one mistake, and the whole operation is done. If we lose a generator, we won't be able to restart it, which means the reclaimer is toast, the Hotsee is fucked. It's a game of chess, and we lost our queen, so we're a couple pawns out here, scrambling to survive long enough to make it to the other side of the board through a sea of white.

Soon after I relieve Micky, Booby Mac relieves me.

I jump into the truck beside Micky, rubbing my hands together like a scheming villain then tuck them between my legs. The feeling of blood returning to the tips of my fingers burns, and I almost miss the numb.

"I don't know about this," Micky says, more concerned than I've ever seen him. "It's too damn cold. It hurts."

We both look at the temperature gauge. It reads -47 Celsius.

Bobby bangs on the window and yells through the glass, "all hands on deck. Shit's going sideways!"

When things go wrong out here, it happens all at once. We live and die by Murphy's Law: everything that can go wrong will go wrong.

The drill's motor starts to whine and sputter. It's gone too cold to run, so it needs to be tarped in to avoid further damage. The generator dies a few minutes later. We have a half-hour window to get it started again, or it'll be done until morning, until it's recovered. The trash pump, used to circulate the mud so it doesn't freeze, quits on us. Like robots, we scramble through the ice and snow to battle the elements.

Micky and Booby Mac start to tarp in the drill, I'm on the generator, and Special Ed is on the trash pump. We all thrive in chaos and, for the first time all night, the boys are smiling, even as the howling winds cut us. We don't care. We finally feel something. We know only that we have to get the job done. No slacking, no bullshit, no excuses.

The wind doesn't matter. The cold doesn't matter. We don't matter. The only things that matter are results.

We're able to get everything working, at least for a bit, and then we jump back into the truck, grateful and hungry for warmth.

"So, you got a place in Ecuador, eh?" Eddy asks me.

"Yeah, man, sure do," I say with a smile.

"Is it warm there?" Eddy asks.

"It's on the equator," I respond.

"Oh, right, right. So, it's warm there?"

"Yeah, Eddy"—I chuckle—"it's warm there."

"Are you from there?"

"Eddy, I'm white and working in Canada. You've known me for months now," I say.

"I was just wondering if you were from there," he says.

"I'm not from Ecuador."

"Oh, okay." His mouth hangs open, perfect little donut. "Me neither," he adds.

That's enough talking for now, Eddy.

It took over ten hours to generate enough warmth for the Hotsee to thaw and start again, but it did. It was going because a bunch of beautiful idiots ripped and clawed its way back to life.

We were back online. The guys had frozen faces and icicles hanging from their noses. Tested from the internal Gods without a word of complaint.

When the shift ended, everything was still running or had returned from the dead. We have our bullshit with the day shift—talk shit, because we have braved worse—then get in the truck to head home and shovel hot food into our mouths.

In the truck, Micky is holding a twelve-pack of Bud Light, chilled to absolute perfection, and wearing a wicked smile on his face.

"Let's just have one, boys," he says. "There may not be a tomorrow."

My Natalia

We have been talking almost everyday. Well, not talking; texting. It's the twenty-first century—people don't talk. We don't really type about our day, because it's the same shit—we work, then sleep, occasionally drink, exercise, masturbate—unless Micky or Booby went a little buck, then we talk about Micky or Booby.

Natalia and I discuss poetry with pain behind it, about the difficulty inherent in life, the trauma we create and exist in, simple wins, harsh losses. We send poems to each other during the night. I read them on the tanks as I work. It helps pass the shift and long, lonely nights.

It's nice when you have someone to talk to on night shift that isn't your typical, oil-stained, mud-covered half-in-the-bag pipeliner. It's nice to have a soft, sweet, artistic lady to break the monotony.

Micky could sense something about me. The way I walked, the cadence of my voice, the spring in my step.

"What's going on with you?" Micky asks. "Who are you talking to? You got this smile pasted on your face. Fucking makes me sick. You're bouncing all about, and it's cold as shit out here. There's nothing to smile about."

"What are you talking about?" I ask with a sly, puzzled look on my face.

"I'm talking about you! You're smiling, and happy!" he shoots back. "You're talking to that girl, aren't you?"

"So? What if I am talking to a girl?"

"You're in love with that girl!" Micky giggles. "Look at you; you're all giddy like a schoolboy."

"Shut up, Micky. You're an idiot. An idiot, I tell ya."

He chuckles behind his cigarette. "Enjoy it now. It'll fade, young man. Trust me; it'll fade."

Micky and Mademoiselle

The Grande Prairie job ended, yet Micky held turmoil in his heart. Things weren't going well for him. He wore it on his face—a hangdog expression, morose eyes.

I want to tell him how much I feel his sadness. How, if I could, I would fix whatever is going on in his head. If it's money, I'll give him all the money in the world. If he's missing home, I'll send him away so he would never come back. Maybe it's something deeper, a pain from childhood, an open wound that never scarred. Whatever it is, I wish I could delve inside and surgically remove it, douse it in kerosene, and burn it dead.

Micky's demon is hungry for something. All it takes is a beer to open the door a crack, and a crack is all it needs to escape.

"Drink up, Micky. We survived another one," I say, tipping my bottle in his direction.

We're sitting in the truck in front of our hotel, listening to music. Micky continues staring at his phone.

"Get the beer in you. It's so good once it hits your lips."

"It sure is. Too good."

"What's up? You haven't been yourself."

"The old lady is driving me nuts. She hasn't messaged me in three days, won't take my calls. Like, what the fuck is going on?"

This is the same shit that happened to Booby Mac. I didn't want to repeat this—watching a friend poison himself. It seems that everything ends in the land of snow and oil, and relationships are no exception.

"You think she's fucking someone?" Micky asks through cigarette smoke. He lights a new one as quickly as he butts out the last.

"What's going on? What's got you thinking that?"

"All these short messages. She's cold when I talk to her," he says.

"You've been gone a long time, Micky. Maybe she's feeling a bit of resentment."

"I know, man. Yeah. But I think she's fucking someone."

"Well, you fuck a lot of women here, right? So, it could be evening out. Karma's a bitch."

"I know. I get that," he says, rubbing his face with his thumb. "I just want to know if she is."

"What would you do if she was?"

"I don't know. But I need to know, either way."

"Fuck, Mick, I don't know if she is or isn't, but they're just like us. If she ain't getting your dick, then she'll find another one," I say and immediately regret that these words came out of my mouth.

Micky's whole body stiffens, a flash of anger present on his face. "Well, fuck you! That's not making me feel any better!"

"I don't want to lie to you … Maybe you should go home. It's your wife, man. This?" I gesture around the area we've called home for the last few weeks. "This isn't real life. There's gotta be more."

"I can't go home. I need to catch up with the bills," Micky says. "She said she wants me home and that the money doesn't matter, but we're almost caught up."

"It's a tough one. I bet she's doing this so you go home. She's sick of you being away, right?"

"I'm going nuts here."

"Fuck it, Micky," I say, handing him a beer.

"Yeah, that's probably the best," he says, staring through the windshield at the lightly falling snow.

"Fuck it."

We continue drinking into the hotel, laying on our beds, driving beer after beer into our bodies, celebrating the end of the job by getting drunk and not talking.

Micky puts back two for every one of mine. Then he switches to liquor, and that gets his demon in a lustful mood.

I see him on his phone, on the back pages, in the escort section. It's uncomfortable being beside a man when his pants are tightening, and his thirst is showing. He needs an approximation of love, even for just a few minutes.

"I'm going to the whorehouse, bud," Micky says.

"I'm coming with you, Micky. It would be irresponsible if I didn't go with you," I reply, fully cut. I swing my legs off the hotel mattress and have to catch myself. "There could be trouble. Hood rats and swamp donkeys are about."

"You're a good friend. There's this Asian place I found on the back pages. We got to go to the house."

"Obviously! But, Micky, the beer. We need more beer!"

"We'll pick some up on the way," he says, dialling a taxi.

Micky gives the driver the address, and he breaks out in a laugh. We're not the first of our kind who he's taken to that address.

"What are you going to do there, my friends?" he asks with a smile.

"Whatever they'll let me." Micky laughs.

"Ah, yeah, we're probably, uh, we are just gonna look. That's it. Just a little look," I say, cracking open another beer. I'm losing track of how many deep I am.

My smile somewhat dies when we arrive at the whorehouse. It's a hovel-type residence with peeling red paint and shuttered windows. Even the bushes around it look like they're on crack.

Out of the taxi, Micky and I stumble to the blank brown door and hear activity emanating from inside. The smell of cinnamon and Eastern spices invite us to knock.

An older woman answers the door and introduces herself as Mademoiselle. What an introduction.

"Come in," she tells us. "Have a seat."

The interior is much nicer than the exterior. The couches are a rich shade of red, and the tables are dark, well-grained wood. The walls are covered in pictures, displaying tasteful nudity. One has a woman wearing a long, beautiful sky-blue dress with one breast showing. She has an apple in her hand, pressed to her teeth, and she's staring at me like I'm intruding on something personal and spiritual.

We sit on the couch and offer Mademoiselle some of our beer, to which she politely declines.

"What are you gentlemen looking for today?" Mademoiselle asks, all business.

"Ah, well, what are you offering?" Micky asks, unsure.

"Blow jobs are one-hundred dollars. Sex for half an hour is one-fifty. An hour is two-hundred. All night is five-hundred dollars."

"What about anal? Micky likes putting it in the butt," I blurt out. A drunk idiot, I am.

"Shut up, I do not."

Mademoiselle smiles. She's heard it all before. "An extra fifty dollars for whatever you want."

I raise my eyebrows at Micky, genuinely surprised at how good of a deal it is.

He responds with a slight headshake. "Can we look at what type of girls you have?"

Mademoiselle brings the girls out, one by one, and marches them sweetly through the room. First, a petite woman with small tits, a flat ass, straight black hair, and a fantastically beautiful face, smiles and waves at us invitingly. Next, a woman in a skin-tight, dark blue dress that ends halfway down her ass. Then a thicker woman who looks like she's an incredible cuddler who would try hard to impress. Followed by another plump one completing the squad from which to pick.

"I like the first two, but the third girl looks like she would break your cock off. But, like, in a good way," I say to Micky.

"I think you're right. She would do a good job," he replies. "But I'm just going to drive 'er in, and I don't think I'll last too long. It's been a while. I'm not here for the journey; I'm here for the result."

"That's a good point." I take a swig of beer and chase it with a shot of Jager. "So, which one, then?" We discuss like we're a panel of judges.

"It's a tough call. They're all so great. There's so much pressure."

"You don't want to fuck this up, Micky. It's a hard choice, but it's yours to make. So, make sure you don't fuck this up!"

The Mademoiselle is looking at us as we discuss. I wonder what's going through her mind. How many times has she seen this play out? Does it even register to her anymore?

"Are you going to get the hour or the all-nighter?" I ask.

"Thirty minutes is all I need. I think I'm going to go with the blue dress."

Mademoiselle smiles, oddly warmly. "Lovely, sir. Come with me."

Mademoiselle leads Micky up rickety steps to a room that is brightly lit and sparse. The only feature is a bed that looks like a gurney. It would be a hinderance to some, but Micky is a trooper.

After a short while, Mademoiselle returns to me on the couch.

"And who would you like?" she asks.

"Oh, I'm good. I got a girl I like back home. I'm going to see her pretty soon, but thank you."

"That's very nice."

"So, what's your real name?" I ask Mademoiselle.

"It's Mademoiselle," she says.

"Cool."

We sit in the agonizing silence of a rural whorehouse while I drink away the sound of squeaking through the walls. No one enters for those twenty minutes as I wait for Micky to nut and clear his mind.

When he comes out, his face is blank and unreadable.

"How was it, Micky?"

"It was all right."

"All right. Let's get the fuck out of here."

As Micky pays Mademoiselle, I can't help laughing when he leaves a tip. It's surreal. I guess this is technically the service industry.

We leave the house, head into the harsh winter cold, and flag down a taxi.

"Be real with me," I say. "How was it?"

"Ah, it was all right," he says with flattened lips and a hint of regret in his voice. "I kind of wish I hadn't done that. I would rather have the hundred-seventy bucks."

"She was hot as fuck, though."

"It's not the same when you pay. It feels cheap."

"Miss the thrill of the hunt?"

"Exactly. It was just a lay down," he says.

"You mean, she star-fished?"

"Kind of. Not really. It's just … vacant."

Micky fell further away than before. After the pain of home, he sought the pleasure of drugs and the promise of an orgasm, but when the moment of sparks and ecstasy drains away, it leaves an even deeper pain. No love, no caring, little less money.

In the morning, we'll wake to a hangover and a new memory of trying something, anything, to replace pain with pleasure and failing again.

Double Day Off

A day off is a splendid thing. I wake up late, roll around in bed, and wait for the motivation to get my beautiful white ass out of bed. There's no clock, and nothing to do except exactly what I want to do. Maybe I'll hit the gym, get swell, then lunch—no, wait. I'll have some brunch, a mimosa to wash down rye toast and eggs benedict. Instead, I lay in bed and read between bouts of tiny naps.

Marcus Aurelius teaches me the ways to enlightenment. I know how to achieve actualization. So, I flip open my laptop and watch a little porn. Now I'm relaxed and mellow. I have actualized.

Natalia arrives in her standard attire. She's wearing tight jeans with holes in the fabric. Not the jeans you buy pre-ripped. Jeans that had no holes in the beginning, but with a little hard work developed authentic, fashionable holes. Her baggy shirt has a llama on it.

"How was your day?" I ask.

"I worked out with my Mom."

"You tell her about me? Does she miss me?" I ask.

Her mother disliked me when we were growing up. She always thought I would end up in jail.

"Oh, she couldn't stop talking about you," Natalia lies.

"Did you tell her I'm not in prison yet?"

"Yet? How long until you are?"

"Depends on how long it takes them to find your body," I say.

"You're going to kill me, are you?"

"I've already bought the shovel."

We head to a bar on Whyte Avenue and start playing pool. She's pretty good at smacking random balls as hard as she can without any intent then having her balls magically fill the holes. They bounce off the shoulders three, four times, then slide heroically into the corner pocket.

"A bet for this game," I announce.

"What's the wager?"

"Sex!"

"But I *want* to have sex!" she says.

"Yes, but sex has many moves, many positions. Dirty, naughty actions that delight the imagination. That's what I mean by sex."

"What's on your sick mind? What's the bet, then?"

I get close to her so only she can hear what I'm going to say. These words are for her ears only. "I'm glad you asked. I would love to finish all over your beautiful face. I just love it. So, if I win, I get to cum all over your face with your reading glasses on, like an innocent librarian."

"Oh. Okay." She pauses. "I can do that. And if I win?"

"What do you want? Me to lick your bum hole?"

Her face scrunches, disgusted. "What? No. No, not that. No."

"You know I could give it a good licking. Get my nose in there. Don't tempt me, Natalia," I say, smiling. "What else do you want?"

"A nice, long massage, top to bottom. And then to sit on your face while you eat my pussy."

"And a little bum lick, too?"

"Shut up," she says, punching my shoulder.

We shake on the terms. I would have done that regardless, but a bet is a bet.

I easily win the game. Maybe she let me. I don't know. Dirty things will happen, wild things, and that's all I care about.

I'm half-drunk when we arrive at the truck. I definitely shouldn't drive.

After a couple tries, I jam the keys in the ignition, and the purr of the V8 rumbles through the air. With Natalia to my right, I put the truck in drive.

I miss being in the country where I can drive the open country roads, free of traffic and, most importantly, cops. Unfortunately, the city is filled with pigs, so like any responsible drunk, I go the speed limit and keep it between the paint. I see nothing wrong with a little drinking and driving. It's a blast.

We head inside my place. Old man Richard is passed out in his chair, an empty bottle of Jim Beam on the table beside him. I get close to make sure he's still breathing. Yup. Shallow, but breathing. Then Natalia and I head to my room.

We put on some tunes, crack open a few beers, lay down on the bed, and chat. The feeling of want becomes greater and greater the longer we chat and lightly kiss.

"Come here," I say.

We get off the bed, and I put her face against the wall. I grab her arms and hold them above her head.

"Keep your arms up and don't fucking move," I

whisper in her ear.

I pull off her shirt. I don't talk. She doesn't talk. I undo the button of her jeans and slide down her pants. She's not wearing any underwear. What a pleasant surprise.

I take her arms down and put them by her hips. I stand behind her, barely touching, feeling the static cling of our skin, and wait.

The air starts to thicken with the silence. We haven't spoken a word. Another minute passes. Another. She can start to feel it. There's no name for it, but it's present. An erotic feeling built from unpredictability.

She can't know what I'm going to do next, because I don't know what I'm going to do next. In these moments, doing nothing heightens the senses and inflames the imagination.

Five minutes pass without a single word spoken before I start to take off my shirt, then my pants, then my underwear. Two naked animals.

I wait a small bit longer before I place my hands on her hips. She presses her ass into my cock. My hands find each of her cheeks. Then I spin her around, and we basically attack each other, lips pressed together, our tongues exploring each other.

She wraps her legs around me as I pin her against the wall. Her hand finds my cock and slides me gracefully into her dripping wet pussy. Thrusting and thrusting, moaning and grunting, I take her from the wall and throw her on the bed.

"I want you on top. I need you to fuck me." My mind is blank. I can't think straight. "Fuck me, Natalia."

She mounts me. What a sight. God, she's beautiful. Her hair falls over her face, so I can only see her eyes and mouth. I pull her in so I can thrust deeper.

"Clean my cock!" I demand.

"You ask for a lot," she says with a twinkle in her eyes.

She crawls backwards, takes my very moist cock, and starts to lick and suck and massage. Then, she stops. She keeps eye contact with me as she moves to her bag and pulls out a small vibrator. What an animal this woman is. Fucking animal!

I flip her over and climb on top. I place the vibrator so it's on her pearl before I drive deep. The combination sends her wild. I find the top speed on the vibrator, and it sounds like a hummingbird. Her body starts to twitch, and her moans grow into yells.

Her arms wrap around me, and I continue, repeating the action again, then one more time. Eventually, the vibrator finds its way to my testicles and pushes me to the point of climax. This is it. The bet was won. The payment was due.

"I have to cum, Nat," I say in a pant. "Put on your glasses."

She does.

"Tell me what you want."

"I want Papa's cum all over my face," she says.

I pull out and finish over her beautiful face, all over those lovely glasses. Eye protection is important for safety. Then I roll over and put my head on her belly, spent.

"Can you get me a towel?" she asks, laughing.

"Right. Right. The mess," I say as I grab her a towel to clean my DNA off her.

"That was intense," she says.

"Yeah," I reply. "All we need is a smoke."

There was another round later in the night, then a third in the morning. The air in the room is distinct. You can smell good sex, breathe it in deeply.

The Boys Take Care of The Boys

Part One

I'm not an angry man, but I can get angry. There is kindness sprinkled in my madness. Insanity layered with compassion. And for my boys, loyalty. There are only a few who can call it their rig. The people who care for their iron, who bleed when it bleeds, who wash her, love her, feed her. Let me warn you: I will come for you if you come on my rig and bark or talk back to the boys or hurt my baby.

A two-hundred-and-seventy pound tub of lard with beady eyes and a double chin, ego twice the size of his belly, walks on site, squeaking about having eleven years of experience.

There are people who tell you they have experience, and there are people who show you. He is class-A dog fucker material. Instantly, I knew our destinies were fated to clash.

I know of Rome. Fearless gladiators, they are my teachers. They are my role models. Perhaps the pipeline God's needed another sacrifice. I don't know what they desire, but I am fated to provide.

I knew him for a mere four days. This man is a big boy. He wears small Harry Potter glasses with a well-trimmed pencil moustache. Red. It's the colour of his coveralls. He's six-four and says anything that comes to his mind.

"What's your name?" I ask him on the first day.

"The name's Boner," he says.

"Boner?"

"Yeah."

"Boner? Your name is Boner?"

"Yes."

"On your birth certificate?"

"What do you mean?" he asks.

"Your parents picked up a little baby, looked at it with love, and said, 'This boy shall be called Boner'?"

"That's my name. What's it to you?" he barks.

"Uh, okay. Nice to meet you. How long you been doing this?"

"Yeah, I got eleven years experience, mostly in downhole drilling rigs," he says.

Downhole rigs drill down to extract the oil. It's the most alpha male environment in oil and gas. I worked on a drilling rig before, and this is the place where you turn sweet kids into men.

"Shit, bro. Drilling rigs. What were you, a driller?"

"No. Just motors," he says.

Big, fat, red flag right there.

Any man who worked in drilling rigs for eleven years and is only a motor hand means he didn't keep a job very long. In five years, you should be the head of the crew as driller. And if, after eleven years, you are just a motor man, then it means you're a failure of some kind.

"Oh, sweet," I say with a hint of skepticism.

"And you?"

"Running tanks on the big job and drilling on the little rigs."

"Ah, nice, yeah. I want to get drilling as soon as possible. I think in a few months I'll be drilling," he says.

Another red flag springs up. It took me two years before I started drilling, and I have a reputation for excellence. It was earned.

*

We were working in a place called Meota in Saskatchewan, close to North Battleford. North Battleford has a reputation as the worst place in Canada. I like it. It has a small town kind of nice to the people here.

We made our new home in a hotel called the Tropical Inn.

Second day on the job, I chat up Micky, the morning shift tank hand, during shift change.

"Micky, how was your day?"

"These guys don't really know what they're doing," he says.

"What do you mean?"

He points down at the workers on the ground. "There's only one guy doing anything!"

"Who's that?"

"Jimmy, the guy with the thick, porn star moustache," he says. "He has a clue."

"What about the other two?"

"I don't know, man," Micky says with a cocked head. "I really don't know. That fat guy in the red, he cleaned the trailer, but that's about it."

"And the other guy?"

"He's kind of a space cadet. Always has an excuse for something. My brother's aunt's dog died; I need to go to the funeral. My ex-wife has a headache; I've got to go home for the week. It might be cancer."

"Sucks to suck, Micky," I say, laughing.

I slap Micky on the shoulder, and he heads to the Tropical Inn to sleep.

It's a big job, so we have two tanks, side by side, and he flanged up the Hotsee to keep the hot water circulating so we'll never have to climb down to get the wash wand. Our tanks are tarped in nice and snug, and it's about twenty-five degrees inside. He left me in pure luxury.

Nothing really happens that night. One of those days where everything goes well. I do my regular exercises. Four-hundred push-ups, four-hundred sit-ups, fifty pull-ups. I have a book of Bukowski poems for when I get bored. In no time, day shift is here.

*

The third day, I return after a short sleep. The transition to night shift is always tough. Your sleep pattern is all over the place.

After arriving on site, Boner comes trotting up to me in clean overalls. He's trying to scan for weakness.

"Hey, there's some garbage over there. I need you to clean it up. I've been working hard all day," he says.

"That's great. Why don't you clean up your mess?"

"I just told you; I've been working all day. Just want everything clean. I've been doing this for eleven years. I like things done a certain way."

"Then do them that way."

"I'll tell you again; I've been working all day. You can do it."

He turns and walks away. I'm taken aback a little, unsure of what he was trying to accomplish, but I'm pretty sure I just got barked at.

Something in my gut starts to stir, anger bubbles. I go see my shift rotation. I go see Micky, who's caked in mud.

"So, what happened?" I ask him.

"Fucking bullshit happened! Those guys"—he waves his arm at the other workers—"those guys are fucking useless. Me n' Porn Star did everything, and they all did nothing."

"I just talked to the fat red cow. He's clean as a whistle," I say.

"Clean as a whistle, yet still full of shit. I had to

run down, do the wire connection, then run back here, do the tanks, almost overfilled it half a dozen times. Barely had time for a smoke."

"You didn't smoke?"

"Thank God I was able to sneak a few, but I'm beat. That fucking guy, calling himself 'Boner,' what the fuck? Tells me he has eleven years of experience, then he tells me he went to jail for three years and has his own concrete business." Micky spits.

"Sounds like he started working when he was ten, eh?"

"Told me he made half a million last year."

I burst out laughing. "Why the fuck is he working here, then?"

"He needs somewhere to spout his bullshit, I guess." Micky shrugs.

"That fat bastard barked at me about the garbage over there. Telling me he worked too hard today."

"He didn't do shit all day. I'll do the garbage before I leave, man."

"Nah, no, no, no. I'll do it. Get some sleep."

I would love to tell you that I am a calm, level-headed guy who has complete control over his emotions. But I'm not.

I would love to tell you that I am a modern-day

Buddha or Socrates, searching for tranquility on a path of enlightenment, a quest for a purer soul. But I'm not.

I'm just not.

In my tanks, I sit and stew. How dare this family-sized bucket of KFC bark at me? He's fucking with our religion, our laws, our rules.

The devil keeps poking at my weakness, my pride, my self-worth, all the times in the past I've been belittled. For hours, I pace back and forth like a wild tiger in captivity. I tell myself I'm overreacting, that this is a waste of energy.

From my tanks, I watch Booby drill. Eddy's filling up everything with diesel and doping rods. Rippin' Ralph and the Cole's are doing wire line.

Shift change. Time to sleep.

That night, I dream that I'm fighting, but the closer my hand moves to the enemy, the slower it goes until my fist is stopped mere inches from them and I can't press further. I dream of frustration.

*

On the fourth day of knowing Boner, I arrive on site slightly after six thirty p.m. and take in the view of all my glorious iron. The drill, the pumps, the tanks, Hotsee, and the flock tank (storage tank that holds mud). I walk along the grounds and see garbage scattered among half-washed tools. The red blimps coveralls are pristine again.

I walk towards the tanks and climb up to visit Micky.

“The fuck happened now?” I ask.

“These guys are an absolute joke. Look at me—I’m covered!” He’s not wrong. There are mud specks blotching his face, and his coveralls are covered in a thick coating of wet sludge.

“Look at all those tools there. You know how I feel about tools?” I say.

“Man, I know. I’m sorry, but I’m doing everything here.”

On shift change, you’re supposed to leave everything perfect. Perfectly clean. Perfectly tidy. No traces of garbage. The site should look brand new, fresh and ready, like a new toy. That’s how every shift change goes, or people start ripping into you for not doing your share.

“Whose fucking tools are those?” I ask.

“It’s that Boner douche. He went to clean them and didn’t put them away.”

"Tell him to do his job, Micky. You're the mud man. Orchestrate your minions."

Exhausted, Micky slowly climbs down from the tank. He wipes mud from his coveralls as he walks over to Boner.

I can't hear shit from where I am, but I can read their body language. Big Red shakes his head repeatedly and angrily before storming away. It starts to boil my blood.

This man is blasphemy. He insults our God's. There is a chain of command out here, and that fat piece of dogshit crossed a line.

I slide down the ladder and storm towards him.

"Hey!" I scream over the diesel motors. He's standing up on the flock tank above me. Others turn and watch the show.

"What?" he says, sounding surprised.

"You get the dicks out of your ears, you whale, and listen. Listen good. There is a chain of command out here, you fat fuck, and you're at the very bottom, you hear? When me or Micky or the driller tell you to do something, you fucking do it. Got that, bitch?"

"What are you on about?" he chirps, still sounding surprised.

My voice can be heard throughout the site,

echoing around the tanks and equipment, rattling around in the heads of everyone. People are running over to see the source of the commotion.

"You get the fuck over there and clean your goddamn mess."

"What you mean? Fuck you. I've been working ele—"

"Eleven years," I cut his sentence short, "and you're still a sack of shit! Micky asked you to clean up your fucking mess, so get your diabetes-filled body over there and do as you're told. Don't fuck the dog on me out here, you prick."

I storm off towards the tool push. Booby and The Chief are there, chatting. They know what's happening. I'm doing my job. Someone broke our rules, spit in our holy water.

The devil has a firm hold of me, whispering his compelling words of delight, making me want to kill. I look over and meet eyes with Booby, who gives the head nod. Finish him …

I see him running my way. I can see through him—weakness covered in rolls of fat. I hate him, what he represents—entitled laziness.

"What's your fucking deal?" he asks, projecting his voice.

"Shut the fuck up when I'm talking to you! Don't talk about your eleven years of I don't give a fuck. Chug rope. You're as soft as a sherbet shit on a hot summer day, you priss."

He puffs out his chest like a gorilla. Probably watched *Planet Earth* one too many times and thought he was part of the wild. He'd die in the wild. He'll perish here.

"I'm the hardest working guy here," he says.

"Then do your fucking job! When we tell you something, you do it. It's that simple."

"I'm going to tell Rossko," he salts back.

I laugh. "You call Rossko right now. Fuck it. I'll call him for you, tell him what kind of dog fucker you are!"

Boner scampers off into the crew truck, but I'm not done. There's still a rage coursing through me, tinting my vision red.

I gather everyone around the drill, day shift and night shift. Booby and The Chief stand behind me. Order needs to be restored.

"You listen here, fuckos. There's a chain of command out here. It goes tool push—that's Booby and The Chief. Then it goes mud man—that's me and Micky—and then there's the rest of you. When one of us tell you to do something, you fucking do it. You work hard, then you'll earn some respect. You don't work hard, you get shit!" I scream. "Boner? Boner is at the bottom now. You got a dirty job, one you can't stand, that you dread all goddamn day? He does it now. He is worse than dirt. You understand? Good. Now get back to work."

Micky comes up to me after the crowd of people disperse. "You got to calm down, bud. That was a little over the top."

I feel complete. The pipeline God's used me as a puppet, and I loved it.

You fuck with our laws, our rules, our religion, we must act with furious wrath and righteous anger and make a sacrifice in honour of the pipeline God's. Amen.

Part Two

"Can you believe that guy?" Boner says.

"Well, what'd you expect?" Micky retorts.

Boner and Micky are at the hotel pub, putting back a couple of cold ones after their shift.

"What's that guy's fucking problem?" Boner asks. "Flipped out over nothing."

"He doesn't like when you look clean and don't do what is asked," Micky explains while rolling his eyes. Shockingly, a man who calls himself Boner ain't too bright.

"Fuck him. I cleaned the tool shed. I've been working my ass off."

"Yeah."

"I've been doing this for eleven years. Eleven years! I don't have to do that grunt work anymore. I put my time in." Boner orders more beers. "Fuck, man. I spent three years in jail; I should have fucked him up. He's lucky I didn't. He's lucky I didn't beat his ass down right then and there."

"Yeah."

"No, I'm serious," Boner says. "He pulled that shit in jail, I would have beat his ass. I never let anyone talk to me like that."

"You did, though."

"What?"

"You did let someone talk to you like that. Today. Everyone saw. Just …" Micky tries to calm the situation. "Listen. He's been here a long time, and he doesn't like when new guys talk back."

"I've been doing this for eleven years. I know what I'm doing."

The boys order more beers. The liquor flows in the veins and does what liquor does best—frees up the emotions and gets people talking without a filter.

Micky turns the conversation to something else when the boys start ordering doubles. Boner takes out his phone and shows Micky a picture of a '70 square body Chevy.

"Shit, son. That's a nice truck," Micky says.

"Yeah, built it myself. Well, with my dad."

"You got any kids?"

Boner smiles. "I've got a little girl. And I miss her."

"I've got a boy and a—"

"She's the best thing that's ever happened to me," Boner interrupts. "I love her so much."

"I get that. Yeah, I miss my—"

"Did I mention I got my own concrete company I do in the summer?" Boner interrupts again.

Micky gives a thin smile. "I thought you worked rigs for … What was it? Ten years? I can't remember exactly."

"Oh, I have. And I do concrete," Boner says. "Me and my uncle do it."

"You and your—"

"Yep!" Again, Micky is cut off by Boner

speaking through his sentence. "We work for the summer. Make a killing. A killing! Like, I have enough money. I don't even need a job. It's just something to do for the winter."

"Yeah," Micky says. "Same with us. We all leave our children to drill in the winter as a fun hobby."

Shortly after two a.m. the bartender shouts for last call, and the guys stock up on beer for the last hour. The conversation returns to me.

"I would fuck that guy up," Boner says with a tight grip on his beer.

"Like I said before—"

"No, I'm serious. I can hold my own. I might look fat, but I'm fast."

"Yeah."

"I don't need to take any shit. I'm not out here to be disrespected. I'm not out here, away from my family, to be around pieces of shit like that."

"Hey, man. That guy's my fri—"

"You don't know how hard it is to be away from your family," Boner says, tears forming at the corners of his eyes.

"The fuck you mean? I have a fam—"

Boner interrupts him with a choking sob. He's trying to hold back tears, sniffing them back, but they're starting to leak and roll down his face. "I love my daughter so much. It's so hard to be away from her."

"Yeah," Micky says. "Uh, I'm out of here. Get some sleep, man."

"I'll come with you," Boner says through a stuffed nose, cleaning tears from his eyes. "Just give me a second."

Micky leaves him at the table in the hotel bar, the only one left, save for the bartender, spilling tears into his Budweiser.

Micky shakes his head and stumbles to his room.

Part Three

Micky and Boner are still drunk on their way to work. Liquid courage still in their veins, unbeknownst to me.

It's nearly quitting time, and I'm tired. I'm up on the side of the rig, standing by a flock tank. I watch Booby drill as I dope some rods. I'm relaxed, leaning up against the tank, eyes half-shut, waiting for day shift to come change us out.

In my mind, I thought Boner would want to have

a little chat. Negotiate, find common ground, and settle somewhere there. I could meet in the middle.

As I relax against the flock tank adjacent to Booby, Boner strides towards me with purpose. He towers over me, his wide, heavy frame casting a shadow. There's a tank to my back, a garbage can to my side, a pump control to my left, and a bright red mound of lard in front of me. I have no exit options. Fight or flight kicks in. Flight isn't the fun choice. I much prefer the fight option.

I square up, arms to my side, chest out, open body position.

"Cocksucker," he yells in my face, shaking his pointer finger an inch from my eyeball. "I didn't come here to take your shit. If you ever disrespect me again, I will fuck you up—"

That's all it took. He threatened me on my rig, my holy ground.

I swing my right fist and connect flush with his left temple. He crumbles like the towers on 9/11.

He staggers to my right, and I run at him hard and fast, and do my best attempt to falcon knee him in the ribs. He buckles back some more, one hand in the dirt, supporting his weight.

"You want to go, fucker?" I yell. "You're gonna

fuck me up, huh?"

Booby comes and separates us as I come to make the kill. He takes Boner away to the crew truck while I pace, full of adrenaline.

"What kind of shitshow you running here?" Boner screams at Booby.

"What the fuck are you doing to my guy?" Booby screams back. "I seen you, little shit. What the fuck did you say to him?"

"Nothing! Dude just went psycho. He attacked me!"

"I know he doesn't do that shit for nothing," Booby says. "You get the fuck away from my guy."

I'm still vibrating, animal instincts coursing through my body.

I head over to chat with the boys to calm my senses, to return to a state of normalcy.

They treat me like a soldier returning home victorious from war, high-fiving me, telling me how jealous they are that I got to be the one to knock some sense into him. The pack approves of what I did.

Booby returns and sees the boys congratulating me.

"You think what he did was great?" Booby says.

The guys go silent, worried that teacher is mad.

"Then you should have seen it with front row seats like I did!" Booby puts his arm around my shoulders and pulls me in tight.

Micky starts shadow boxing me, imitating being Boner getting knocked around.

"What he say to you?" The Chief asks.

"He had his finger in my face, telling me how I disrespected him, how he's going to fuck me up if I do it again." I laugh. "So, I disrespected him again."

"You're a fucking beauty," Booby says.

The aftermath comes quickly, like it always does. I'm truly happy I hit the fucker, but now a call from management will come, suspensions likely to follow. Sure enough, less than ten minutes after my punch landed, Rossko calls.

"Hello," I say, overly friendly.

"What the hell is going on out there?" Rossko says sternly. "That Bryan guy called me and said you just punched him in the skull?"

Boner's name is Bryan? And he goes by Boner? What a dildo.

"I'll tell you the truth, Rossko. I did punch him in the skull."

"Now, why did you do that?"

"That fucker threatened me. Said if I disrespected

him again, he would fuck me up."

"That's not what I heard. He said he just came to talk with you, and you hit him."

"That's bullshit, Rossko. You know I don't bullshit. I was cornered between the flock tank and a bunch of shit when that big bastard comes running up, telling me he'll fuck me up. What would you do?"

"To be honest, I would knock the prick out," he says. "But that's not the point right now. Listen, you can't be punching people out on site, all right?"

"Am I going on suspension?" I ask with my tail between my legs.

"Hell no! What? No, no, no," he says. "This is what you're going to do. Don't do anything now. Let him cool off. Then, next shift, go and apologize. Shake his hand and smooth things over."

"I don't think he deserves an apology, Rossko."

"Hey! Stop fucking around! We got a job that needs to get done, all right? So, do it for me." He hangs up.

Rossko is like a father to us all, and sometimes a father needs to give the rod so as to not spoil the child. He's the foundation of our company for a reason.

Calls start rolling in from Rory the mechanic,

Chiggins, few of the other boys on other crews, even Wilton heard about it within the hour.

Men are a bunch of wild animals. When they hear blood has been spilled, they get all fired up, frothing like sharks when there's a blood bath of sailors battling pirates in the dead sea just off the shores of St. Lucia.

Part Four

After centuries of blood baths and battles, murders and rapes, through stealing and larceny, society decided to make rules, then codified these rules into law. If you break these laws (and get caught), then there needs to be a penalty, some repercussion strong enough to act as a deterrent. Previously, society would put lawbreakers in a hot box, or tar and feather and parade them through the streets, or hang them from a tree in the public square. The acts of lawbreaking were documented and recorded so society would know who made a slip-up and who was a real bad egg.

In the laws of nature, the strong beat the weak and assert a sense of order.

In the laws of society, I hit someone and he

didn't hit me back. That's a problem in the laws of society, because it means I assaulted him. Out in the bush, in the freezing cold, in a world of men, we make our own rules. We take care of our own.

I wait for breakfast with Boner on my mind when I get a message from a worker on day shift.

Boner went to the cops.

I immediately call Rossko.

"Rossko, what the fuck did I hear? Did that guy go to the cops?"

"Yeah, he did. I was just about to call you. I need you to go to the North Battleford station after work and write a statement about what happened."

"Are you serious?"

"Yeah, I am. Just wait a minute so I can get in my office," he says. He hangs up.

I'm flushed with stress, thinking about how I won't be able to travel anymore, how my dream could be ruined by a grown man who calls himself 'Boner.'

My food arrives, and I can't eat a bite.

Rossko calls back.

"All right, my door is closed. So, this is what you're going to do: go around to all the boys tonight and get statements off anyone who was there that saw this guy was really threatening you, that you were backed up in a corner. He was towering over you, and it was self-defence. What will happen is this Bryan guy will look like he's lying. Get on it and call me back when you're done with the police. And you didn't hear this from me."

"Okay," I say. What more could I say? It was my only option.

Around the table, the boys are eating, relaxed, enjoying talking about nothing. I tell them I need a favour.

"What?" Booby spits out his waffle. "He went to the cops?"

Rippin' Ralph laughs deeply, head tilted back. He laughs so hard it appears painful. He tries to say something but can't bring himself to stop chuckling.

"Over one little punch?" the Cole's say in his high-voice.

"Yeah, he did. I need your help," I say humbly. "I have to write a statement after work … and I need you guys to write one in my favour, saying he came in threatening me."

"Sure thing," says Special Ed.

Rippin' Ralph laughs even harder at the request.

"Oh yeah, fuck that guy," Booby says. "We'll spin this up and make him look like a psycho."

"What has happened to people nowadays?" the Cole's say in his low voice.

"This is amazing. Yeah, I'll help," Ralph says, wiping tears from his eyes.

"Who goes to the cops over one little punch?" says the Cole's. "Take it like a man, get up, go on with your life."

We all discuss how we are going to write this statement up, describing how this Boner character was threatening me. We align our stories then summon for some pens and paper and begin drafting it at the breakfast table.

We arrive on site and see that Boner didn't work his shift. Too shook up from one little punch after three years in jail, I guess. Unbelievable.

I already have four statements before Micky and Porn Star offer to write one. Clearly, Boner made enemies, because people are lining up to lend their voice for my side of the story. Even the pipeline boys not from our crew offered fake statements, but I already have seven statements in my back pocket, all in my favour.

The laws of nature are far stronger than the laws of man. Respect, honour, hard work, virtue—the attributes we use to quantify strength—were lacking in Boner.

Shift change, and we start working. My mind races like a horse around a muddy track, around and around, replaying what I'm going to write. I'm stressed.

That fat fuck, that little bitch. Who calls the cops? A rat does. A filthy, dirty, fat fucking rat. That fat rat could destroy my whole Ecuador operation.

I'm pacing up in the tanks, waiting for a moment I can't control. I'll kill him; that's what I'll do. After the statement or before? I don't know yet. Maybe I won't. I'll pace some more and think about it.

The night turns to morning, and I'm exhausted and hungry, been too stressed to eat all night.

I arrive at the cop station in North Battleford.

"It'll be all right, man," Booby reassures me. "You got seven statements all saying the guy was drunk and threatening you."

We enter the station and hear the drunk tank regulars talking in circles, mostly to themselves, yet as part of a group.

Booby puts a comforting hand on my shoulder. "You'll be okay," he says then slightly pushes me to the front desk.

"I'm here to write a statement for the Boner file."

“The what?” the clerk asks.

“Right. The Bryan file. About a fight at work.”

“Oh, good. I’m glad you came in,” the clerk says. “I’ll be right with you.”

Booby, Special Ed, Rippin’ Ralph, the Cole’s, and I sit in the station. There are statements in my hand from Micky and Porn Star. We’re a united front when an officer comes out to us from the back.

“Good morning, gentlemen,” he says. “I’m going to need you to come back and write a statement about what happened.”

“Officer, I have a few statements from the other crew,” I say.

“What about?”

“How the guy was wasted before coming to work. One guy said he drank with him until closing. Another said he reeked of booze when he came to site.”

The officer nods as I speak, but I can’t tell if he’s believing my words. “Okay, give them to me,” he says.

I pull them from my back pocket and hand them over.

“So, who here saw what happened?”

In unison, the boys speak, “I did.”

"If I could get statements from all of you, then that would be great," he says.

He brings us back, one by one, to write our prepared statements. The boys write out statements, one after the other, saying the same story with a few small differences. The boys have my back. They write and twist the story to make me an unwilling victim, fighting for my life. Hell, some didn't even see what happened, yet here they are, in the early morning, lying for me. The boys take care of each other.

I shake hands with the cop before we leave. I feel a weight melt from my shoulders.

In the truck, I call Rossko, who answers immediately.

"You go to the cop station?"

"Yup."

"Boys write a statement?"

"Yup. Seven of them."

"Good," Rossko says. "Now this is what's going to happen. You'll get a call in a few weeks, and it'll be a wash. They're going to throw it right in the garbage."

"How do you know that?"

"I know," he says with absolute confidence. "You're not the first little spit-fuck to knock out a coworker on the patch."

"Thanks, Rossko."

*

Two weeks later, I get a call from the North Battleford station and, sure as Rossko predicted, it was a wash. It was like nothing ever happened in the eyes of the law, but it grew into one of the great stories told within in the company, about how this little fucker knocked out this big fat bastard. The story has morphed beautifully into a myth, growing taller and taller with new details in each retelling, additional moments and pieces that never happened, but I never correct them. I let myself become myth. I watch it become pipeline legend.

Booby's Divorce

Booby's divorce is in full swing. I mean, we're nearing D-Day. It's dirty warfare. A battle of emotions and money and he-said, she-said, smoking guns, who did what with whom type combat.

My Booby, my dear friend, is in pain every goddamn day. He drinks a lot now, even by the standards out here. He's started doing blow and, perhaps related, has turned into a real dickhead. Brother Booby was getting a divorce, which meant I was part of it as well.

I spent almost every moment of each and every day with Booby, and I enabled each line of hate talk, murder plans, stupid bitch this, losing money that, angry babbling, punching truck windows. And I was there for each messy, hungover morning. Throughout the weeks, I, too, grew angry and sad and mean. We started to battle.

"Hey, little bitch," Booby says, his tone dripping with disrespect. "What you fucking doing?"

This is how it started. At the beginning, I did

nothing. I knew he was unhappy, and I let it go, told myself it wasn't a big deal. But you can't give an inch out here, not around pipeline men. Every day, he would call me names, whatever it took to get a reaction. His wife was winning this divorce, so he needed to be a man again, somewhere, somehow.

"Grab me the wrench, bitch," he says in front of the inspector.

"What'd you just say to me?"

"You heard me, little bitch."

"What's that? You want daddy to give you a spanking?" I tackle him to the ground.

To the inspector, it looks like a fun fight, but it's not. I get on top of him and start punching the ground beside his face.

"Huh, look at you, you weak, pathetic, little man," I say. "You've got fifty pounds on me and can't do shit."

I put my glove in his face. He opens his mouth and bites my finger.

"I'll rip your finger off like you owe me a gambling debt," he says. "That's how I'll beat you. I'll let you punch my face til you get tired."

"Wanna try it?"

I stand and help him to his feet, but there's

tension. Someone is going to need to be a victor by the end of this.

Later, in the truck, he puts his hand in my face, not touching me, just hovering it to the side of my vision like a fly buzzing around my head. I can't ignore it. It's like he's killing me by a thousand tiny cuts. Alone, each nick is nothing, but added together, it's bleeding me dry. No wonder he's getting divorced.

He starts ranting about his soon-to-be ex-wife. They're fighting over the couch, the bed, over a set of knives they got for Christmas from a friend, how she's set to get spousal alimony because her quality of life isn't allowed to change. Her divorce attorney is trying to find out how much Booby's going to have to pay to make this divorce happen. It's looking in the realm of $60,000, plus the furniture and other items that she claims she's too emotionally attached to. All this is on his mind, and his hand is still in my face.

"Skippy, what's wrong? What? You don't like when I put my hand in your face? Don't like when I don't touch you? You want me touching you, don't you, fag?"

"Who do you think is fucking your wife right now?" I say.

"What?"

"Like, how much cock is she milking right now? She is still your wife, right? So, you're being cucked right now, at this moment, while you wave your stupid hand in my face."

"Probably lots of cock … I don't know," he says in a saddened voice.

"And how much pussy you fucking? Nothing? Each night, you take a solo trip to Palm Springs, don't you?"

"Fuck you."

"Fuck yourself; it's cheaper," I say. "You're a sad case, man. Sad fucking case."

Not long after that, Booby is on every dating site on the web. Booby's a local on every girl's Tinder, Bumble, Match, POF, Hitch—you name it. He is a finger-swiping fiend. He was in a relationship for seven years, and now the cat's out of the bag.

He found a girl in Vernon, six-hundred kilometres away, who was as desperate as him. She's paper bag unattractive, yet he drove the entire way on our two days off just for a lay.

For the first shag, he lasted about twenty-two seconds. But he started practicing, putting in the work, grinding out better performances like he was training for something. This was a new Booby: a hate-fucking, money-spending, mean-mugging, beer-chugging son of a bitch.

We stopped calling each other by our names, so now he calls me bitch, and I call him fag. He would get drunk every night. I sleep and work out. We'd go out of our way to avoid the other, except that we shared a hotel room. Oddly, as soon as we entered the hotel room, we went neutral. It was like our lodging was Switzerland. We would be civil, order food, stay on our side of the room, quiet, relaxing, and neutral. Of course, Booby was still on the bender pretty hard, cursing his wife for throwing parties on what was rapidly, and legally, becoming not his house and travelling, spending his money that he had to give her. But the room, for the most part, was sanctuary.

Wild Night for Booby

Booby's old lady was throwing another party at his house. I don't begrudge a woman for having a party. It means nothing to me. But Booby decided this is his line in the sand and expects me to be on his side. It was a point of no return for him.

Obviously, she didn't listen and was having whole townhouse shakers. Booby found out from a friend and went off the rails.

From across the hotel room, Booby spouts off about how he's going to kill her … blah, blah, blah. She's throwing a party … blah, blah, blah.

"What are you going to do about it?" I ask.

"I can't do anything. I'm here."

"That's coward talk. Are you a coward? Are you a bitch? Go out and party here. Go out and fuck more people than she could ever manage. Take back your manhood."

"Yeah, but—"

"Or don't, fag. I don't care," I say and resume scrolling on my phone.

He stands like he wants a fight, but then quickly relaxes. He stares at himself in the mirror, beside the television, and I watch him out the corner of my eye.

Still angry, he storms to the bathroom to have a grumpy shower. I guess Booby is going to the bar.

After enough rum and Coke's to fill a pool, plus some pure, uncut powder from a girl who took a shine to him immediately, rubbing her fingers up and down his inner thigh until he agreed to leave with her, Booby went back to an amazing gold-plated house. He was riding a perfect line, the kind of dirtbag legends, where the coke sobered him into a suave, debonair gentleman. The name is Mac; Booby Mac.

The house spares no expense. There are trained monkeys hanging from crystal chandeliers, pouring Cristal into fine, tall champagne glasses. Tuba-playing elephants, gymnastic sloths.

The women, too, were imported from all over the world, searching for the best. Barefoot Japanese women in silk kimonos; thick, sultry women from Colombia; and lavish blondes from Sweden with perfect dimples.

The party raged on until morning and, as each new hour came and went, the women took off a piece of clothing until it gradually became a wild, animalistic orgy. Cocks thrusting deep into finely shaved pussies, the smell of cinnamon, wine, and sex. There's a machine in the corner that sprays the air with a mist comprised of cocaine and Viagra to keep the men running hot. This is a carnival. This is what Booby needed.

But, of course, that isn't what happened to Booby. He woke up on the floor of a random house with peeling paint, second hand moth eaten couch, and a couple of back alley lamps. Only him, alone, on the cold, barren floor, no blanket, pillows, or answers. A sad, half-naked blob of destructive flesh, rotten through and through.

He missed work for the first time in his career. Beside him on the table is a pile of cocaine. He remembers nothing.

He racks some lines to give him the confidence to call a cab. This is a true walk of shame. The boys are pissed. His bank account says he's spent seven-hundred dollars.

He rolls himself into his hotel bed and wishes his breath would fade, but he keeps living, depressed, sad, and broken.

Life isn't perfect for the mad.

The First Rays of Hope

The first rays of the sun feel like Jesus blessing you personally. A warm spotlight from some burning ball of gas a bajillion miles away feels like the heavens parted and kissed you on the forehead. It's the beginning of the end. Breakup is coming. After a cold, brutal, frosty bitch of a winter, spring has sprung.

It's the start of March, and the snow has begun to melt. It starts slowly, but in short order, this place will be a mud-fest with road bans on hard and heavy. Production will cease, and the boys will go home for much-needed rest and relaxation. The end is in sight.

"You can feel it. Can't you, fag?" I say.

"Yes, I can, my little bitch," Booby says.

"It won't be long now. Breakup is here. I can't wait for this season to be done."

"I could keep going. Fuck knows I could use the money."

"You'll be all right, bud. You can rip the shop, and they always have those mad scramble jobs at the end of season," I say. "There's always work somewhere for a guy willing to do it."

"Yeah, it'll be all good."

"You know what's funny? This is your second breakup of the year. You like this one better than the last?"

Booby comes and hugs me. "You're my stupid little bitch, ain't you?"

"I'm your daddy, but you can call me Papa."

Every day, the roads go from a white sheet of ice to a light brown, and the frost seeps into the dirt and turns it to mud. Six months of God blowing huge, white loads all over us now had its chance to escape into the ground.

Around mid-March, the ground is still frozen, but when we start to leave site to head home, the top layer turns into a derby. Each truck is in four-wheel drive, pinning the pedal to the floor, and shooting big chunks of mud and rock at the vehicles behind. Each vehicle is covered with a thick sheet of grime and rocks chip away at the windshields and exterior paint. It's an unenjoyable time of year, carrying ten pounds of mud on your boots and overalls as you walk.

This time of year, the equipment is beaten-up and ready to give. It's been in the field for the last three months, same as us, and it's feeling the wear and tear. Plus, people get the feeling they'll be home soon and get a little squirrelly, little less smooth in their movements with the machines.

"I need you to fix the fingers again," Booby says.

"God, everything is broken on this bitch. The top drive is getting pretty hot, too," I say.

"How hot we talking?"

"I can't touch it type hot."

"Shit," he says, sighing. He gets out and feels the top drive. It's not good. We're pulling pipe, though, so we can't stop. He shakes his head in sad frustration. "Whatever. Fuck it. Keep rolling. Run it until it blows."

An hour later, our ears are assaulted by the grinding and screeching of the gears. Then the inevitable hiss and boom. Dead in the water. Game over. Shut 'er down.

Booby doesn't seem to care much. With all this sun on our faces, a feeling of warm permeates our body and spirits, and the long hours of daylight revive our spirit after the hellscape that is a prairie winter. People are nicer to each other, drinking is a release instead of a medicine, and the days pass by easily and without stress, yet this is when we go home instead of work. Isn't that strange?

The End is Nigh

I want to leave in a deep and desperate way. I'm consumed by an anxious need to pack up all our shit and hit the road. Work is over, baby! I'm going home, and I'm never coming back!

It's always a goddamn nightmare packing up broken shit in a mad dash, throwing all our crap in the trailer, getting bills signed. This hectic season is behind me.

When I arrive at the shop in Lloydminster, the boys are there to greet me.

"So, you're out of here, eh?" Booby asks, a beer in hand. There's a spring breeze blowing, and the smell of damp melt mixes with fertilizer. "What am I going to do without you?"

"Everything will probably turn to shit. I imagine you'll be dead two weeks into next season without Papa around," I say.

We clink our beer cans.

"Wouldn't that be nice?" he says.

"It's okay, son. You can call any time."

“So, like, what are you going to actually do?” Micky asks.

“I’m going to get water to the land, then electricity. Make it an actual place I can live,” I say.

“No hookers?”

“No time.”

“Nah, there’s time. You’re in love with a girl, ain’t ya? You’re a good boy now.”

“Shut the fuck up, Micky,” I say.

“Yeah, you’re in love.”

“Is your little honey coming down to visit?” Chiggins asks.

“Don’t think so. She’s not even my girlfriend,” I say. “What about you, Micky?”

He finishes his first beer, shakes the last drops into his mouth, and opens another. “I’ll go home for a few weeks, see the wife, raise my kids for ten, twenty minutes, and then get back to work.”

“That sounds all right,” Booby says.

Micky laughs. “It’s all I know.”

“And you, Booby?”

“Well, I’m broke.” He laughs.

“Perfect! Just the way I like you.”

“Like I was saying, I’ll take a week off then get back at ’er. I’ve got a Tinderella I’m going to have to chase down."

"You're a romantic, Booby," I say.

He nods as he chugs his beer.

"Chiggins?"

He mumbles out, "I don't know. Probably just chill out."

"Try not to wreck your nose when I'm gone," I say.

"Shut up. I don't even do that much now."

"I heard you did it last night."

"Yeah, but, other than that."

"I'll find you a guy in Ecuador so you can visit," I say while embracing him.

Inside the office is the big guy himself. Rossko is wearing his bright green Roughriders hat behind his desk.

"What are you doing here?" he asks.

"Just came in to thank you for a great season before I head to Ecuador."

"How nice. Well, it was great having you. You'll always have a job with us."

"Thanks, Rossko."

"I mean it, kid," he says.

"I really want to thank you for what you did for me with the police."

"We take care of our own out here. That guy is gone, and you're here in my office. There's a reason for that," he says.

We shake hands, and then I walk out the door.

I give each of the boys a big hug as I say my goodbyes. Then I hop in my truck, rev the engine, crank the tunes, and drive into the sunset, towards home.

As I drive, I feel the earth turn to the beat of the universe. A wild, rhythmic chaos swirls around me in alternates and mixtures of madness and beauty. My future is mine to create, my fate my own to write, but my past is in the pipeline, my blood and sweat and tears run through my truck's engine. I'm grateful for another season in the pipeline. When I say, "It could have been worse," you'll know I'm speaking from experience. There is no good and evil, only the strong and weak, only pleasure and pain.

Epilogue
Booby

What happened to Booby? Was there a happy ending? Of course not; they don't exist.

Booby was put through the wringer. One season of getting scraped through glass shards seemed like a lifetime. Divorce is a cruel bitch that broke this man in the mind and in the bank!

Pain? You learn best through it; grow the most while it holds your head below water. It's like taking steroids. Booby transformed into a hate-fucking, mean-mugging, beer-chugging machine, a new man released into the wild. You ask anyone, this is the first stage after any divorce.

"Hey, you little cocksucker. What's going on?" Booby asks through the phone.

"Booby, ya lunatic. How are you?" I answer back.

"One sec. Let me connect the call with Micky."

"Hello?" Micky says after a moment of pause.

"One more, boys. I just have to get Wilton on the

line."

"Harrrrd, harrd, hard, what's going on boys?" Wilton shouts.

"Well, guess what I've got here, boys?" Booby says.

"Tell me!" I say

"Grab some beers, get the fireworks ready."

"Woo, gimme a second," Wilton yells into the speakerphone. I can hear his feet pound the stairs in a sprint. He returns, breathing a bit heaver. "Okay, I got a beer. Why am I drinking?"

"Someone give me a drum roll," Booby says.

The boys pound their hands and stomp their feet in a clueless rhythm.

Booby screams into the phone, "I'm officially free from that bitch! I'm divorced!"

"That's what I'm talking about!" I shout.

"Finally." Micky laughs.

All you can hear from Wilton is the sound of him chugging his beer.

"And that's not all," Booby says.

"What else could there be?"

"Oh, hell yes, there's more," Wilton clamours for more to celebrate. The pop of another beer opening hisses through the phone.

"Well, I kicked that bitch out of the house, right? And as the last *fuck you* to me, guess what she left behind?"

"A positive test that's she pregnant with your child!" Micky spits.

"Daddy Booby," Wilton says.

"Oh shit, I'm a grampa now," I say.

"No, even better. She left a picture of the guy she was fucking pinned on the dart board."

We all burst out laughing, matching Booby's volume. Barking, booming laughs punctuated by almost childish giggles.

"Oh man," I say through giggles. "You did get cuckolded!"

"Yeah, but I'll get him," Booby says.

"Do you know the guy?" Micky asks.

"No, but I've seen him around town. And when I get the chance, I'm going to beat his ass into the pavement."

"Nah, burn his eye out with a cigarette," Wilton says casually.

"So, now what, Booby?" Micky asks. "You're a free man to do what you please."

"I've got a simple plan, boys," he says. "Go to work during the day, swipe right to pass the hours, find a girl for the night, and bang her brains out. Repeat."

"That doesn't sound too bad," Micky says.

"I'm a changed man! The only thing to expect from me is the unexpected."

"The road of madness is a wild one, Booby," I say. "Should I be worried?"

Booby gives a purr, "Oh yeah. Everyone should be ready, because the cat is back. And on the prowl!"

I talked to Booby over the summer, from time to time. The man changed. Every time I called, usually around noon, he was just getting up. During our chats, he would go to the toilet and throw up what was left rotten in his stomach from the night before. He'd explain he was in some random town with some new Tinderella that he fucked the shit out of and how he had another lined up that night in a different town. He was making up for lost time, making notches in his old leather belt. He was on a beautiful road of destruction, a train with no brakes, but the healing process had begun and, eventually, after enough beer had been drank and enough pussy had been fucked, he'd run out of steam. When will that be? Who the hell knows?

Shout Out to The Boys

What happened to all the boys? Well, first off, there are the ones who didn't make the cut, who went home with their tail between their legs, who no one remembers. There's not one person, other than their mom, maybe, who cares if they're six feet beneath the black top, because they're cowards, and everyone knows God hates a coward. But the ones who stayed, the beautiful, destructive creatures I call my brothers, the lunatics who labour in the shadows, who drive long necks and breathe in short lines. They get a full twenty-one gun salute!

I know none of us are painted as smart. We definitely lack some of the decency of any white-collar, tie-wearing snob. Money management is defined by our industry. We're boom and bust! However, when all of us junkyard dogs put our brains together, by sheer will, we can accomplish anything. If a mountain is in our way, we remove the mountain. We can't ford the river, so we go under it. Goddamn, if a valley thinks it can slow us down, we'll fill in the craters until it looks like the prairies!

Micky paws at his face in the airplane while he stares out the window. It's been two hours since his last smoke. It was a long, hard, and not very sober winter, but he was almost home.

As the plane descends, he worries about the transition he'll need to make between the two sides of himself. The wild, liquor-loving madman can't exist when he needs to be the loving father he also is.

God, I could use a fucking smoke, he thinks. He picks at his jeans and searches for loose threads in his clothes. The anticipation is killing him. Will his wife still love him? Will his kid recognize him? It's been so long.

He walks towards the doors, *EXIT* marked above them in bright, ominous red.

The doors slide open and the sudden light momentarily blinds him. He scans the terminal. *All these Newfies ... Who are they waiting for? Where is my family?*

A familiar voice hits his ears from the crowd, though he can't see the source. "Dad!"

Micky's legs are hit from the right by a hug from his son.

"Mom said we were going to pick up Grampa!"

"No, it's me. I wanted to surprise you."

"I can't believe you're here," he says. "Are you here to stay?"

"Yeah, I'm here now. Why you crying?" Micky asks, though his eyes are mimicking his son's.

"I'm so happy you're here. I can't believe you're here," he says again. "I missed you so much."

His son buries his face into Micky's chest, eyes leaking.

Micky's crying, his wife is crying, everyone is crying, and it's beautiful.

Micky looks up at his wife and smiles, the most inviting, evil, corrupting smile that only Micky knows how to do. All is forgiven. He pulls her in and gives her a nice big fat kiss on the lips.

The other boys burn through cash like they hate it, never wanted it. Chiggins does what Chiggins does best—literally blows cash up his nose, drives around with a twenty-four of Pilsner on country roads, and somehow finds the craziest women who have an appetite for cocoa puffs. A man of simple taste!

Wilton has an eye for the shiny and new! If it sparkles, gleams, or has bigger tires than are legal, he buys it. He took his new truck to the dealership and traded it in, plunked his ass into a newer, bigger, better truck. Now that he's cruising with a platinum, he took a little trip to Best Buy and found 4K and sixty-two inches of visual poetry and threw it in the back. Sun was a little bright, so he stopped by Ray-Ban and left looking like a Blues Brother. Another day well spent while spending, he kicked back into his recliner and popped open a beer to watch his new TV. Somehow, he misses work, but he can wait.

As for P-Unit … Do you care?

All of us will be back in the patch. We'll always be back. Alberta is forever.

Ecuador

Corruption is a beautiful thing. The third world system is more personal, more about who you know. It differs from the first world that is colder and more calculated.

Get in line, grab a number, purchase a product. Whether you pay more or less, the permit is coming at the same rate.

But, in the third world, it pays to make friends or be able to provide favours. Barring that you need some bribe money. Ecuador is corrupt, but it can work for you.

We are negotiating what I could pay for a permit. Notice I didn't say, "we are negotiating about what the permit costs?" I could pay what normal Ecuadorian's pay to get their water rights, but that might take six months. And they know I'm white and, to them, white means money.

The permit officer knows the game well. He's played it many, many more times than I have.

He tells me he doesn't know if he can get the

water permit, blah, blah, blah. I wish I could fast-forward through these parts—the faux rule of law—but Ecuador is teaching me patience.

Pedro tells me that we'll need to bribe the man. It costs me the equivalent of one-hundred dollars to get the water permit today. This is nothing, but it's important to always negotiate down. If they know you're weak, the charges will continue to rise. Corruption is a delicate game, but it's fun to learn.

We bring the man down to eighty dollars and get our water rights. Small victory.

The next step is letting the neighbours know what's happening, because pissing off the neighbours can cause all types of problems. You upset the wrong neighbour, and suddenly blockades are set or machetes are pulled on your work crew. The neighbour atop my hill is called "El Doctor" by the locals, but I can't imagine he's a physician of any kind. Maybe a surgeon, in the right environment. He's building a huge hotel about three-hundred metres from my land, and the word around the campfire is that it's to launder money.

As we build our water pipes to my land, we let El Doctor know that we will be going by his land and digging by the pipeline that is adjacent to both our lots.

He warns me about Ecuadorian oil. You don't fuck with Ecuadorian oil. The oil is government-run and the military is quick to act when the pipelines are badgered. We're trying some shady work.

I ask Pedro in my broken Spanish, "How do we get across pipeline?"

He explains, "we dig shallow trench. No one will know about what we do. Everyone does it. Don't worry!"

"Okay, risky business, me gusta," I say with a glint in my eye.

Within the hour, we are caught with our dicks in our hand when a military convoy armed with AK-47's rolls in.

The Commandante strolls up with complete confidence over the situation. "What are you doing here? This is the land and property of Ecuador!"

Pedro tells the boy to hide me, so they don't thing I'm running the show. Things could get a little pricey.

Pedro plays possum and explains, "we are running a water line and digging a shallow trench to my land. Are we doing something wrong? Did a neighbour call?"

The Commandante signals his troops with a hand gesture, and they all tighten their grips on the sub machine guns. "This is a routine check!" His soldiers come a little closer as he lights a smoke and stares in Pedro's eyes. "You need a permit to do any digging on any Ecuador pipeline roads."

Pedro explains, "Is there anything we can do today that will let us be able to get this pipe in today?" Thinking a bribe would smooth out this situation.

The Commandante takes a long, slow inhale of his cigarette then blows it out. "If we catch you here again without a permit, all of you will go to jail. Entientes!"

Three days later, El Doctor calls and tells us that he's putting in a water line for his hotel and that we can join him in getting a water line to my land. How convenient.

This is a bad idea, as he would then have complete control over my water situation. You don't become El Doctor without knowing how to gain power over people.

I humbly decline his offer.

It took time and money, but we got the permit. What we will do is directional drill under the pipeline.

The boys came in with all their equipment. It was like nothing I've ever seen. They had an inflatable kids pool as a water tank, the drill was covered in rust, tape held parts together, oil was leaking everywhere. It was like looking in a trick mirror but instead of white dirtbags it was Latino dirtbags—same shit, different county. But if it works, it works, and we got it drilled to the other side. The water line is working.

Electricity was next. This was pretty simple! Basically, you need to put some poles in the ground and run a wire to each pole. Ecuador is on Latino time, and if it can't be done today, do it tomorrow. So, after many tomorrows, we had electricity.

Things started to get a little more complicated as we moved on to the next steps. Again, I told you how I've been winging it my whole life and, usually, it works out. You throw enough energy at something, and it will yield. The problem is my inexperience and not throwing the energy or money in the right direction! The learning curve is steep.

After the I-can-do-it-myself approach, I get help. I talked to Gringo's who have built houses in the area, talked to people who had trusted contractors, and I found an architect. Most importantly, I found out how much I didn't know, something I really enjoyed. I was a Special Ed, asking all the dumb questions, but I learned. Thank God.

We started talking shop. How are we going to do this?

First, I needed a plan. Those are important! Plans with as many details as possible.

How many houses you want to put up? Where is the best location on the land to put up these houses? What design will the houses be? Will this be rental or personal? Will I need retaining walls? The list went on and on.

Before anything could start, we needed to survey the land to know what was even possible. The property is on a hill, which means we need to make cuts in the land for the foundation of a house. You only get one chance to do this, or you're spending money on a pacer if you gouge too deep.

After the dimensions were placed into a computer and the info was gathered, the architect and I began drawing up plans, kicking around ideas. In the end, we decided on two houses and made drawings and created dimensions so we could use vegetation as a retaining wall as part of the design of my pimp crib where I could roam around shirtless with my golden gun.

So, after the wasted money on the do-it-yourself project and the well-spent money on the do-it-the-right-way version of the project, we got a grand total of what the project would cost. And, well, I didn't have enough. Whether you go learn at a university or learn life's lessons by trial and error, both cost money. I didn't have enough!

It made no sense to build something halfway. Materials could get stolen. Plans could change. An earthquake could crack the very foundation of my dreams.

I failed to meet my mark. I needed more money. This is what I feared. Luckily, I know a place where I can get it … The place of the mad, the place of the wild.

The pipeline!

www.ingramcontent.com/pod-product-compliance
Lightning Source LLC
Chambersburg PA
CBHW030607220526
45463CB00004B/1204
9798662967922